Lecture Notes in Networks and Systems

Volume 44

Series editor

Janusz Kacprzyk, Polish Academy of Sciences, Warsaw, Poland
e-mail: kacprzyk@ibspan.waw.pl

The series "Lecture Notes in Networks and Systems" publishes the latest developments in Networks and Systems—quickly, informally and with high quality. Original research reported in proceedings and post-proceedings represents the core of LNNS.

Volumes published in LNNS embrace all aspects and subfields of, as well as new challenges in, Networks and Systems.

The series contains proceedings and edited volumes in systems and networks, spanning the areas of Cyber-Physical Systems, Autonomous Systems, Sensor Networks, Control Systems, Energy Systems, Automotive Systems, Biological Systems, Vehicular Networking and Connected Vehicles, Aerospace Systems, Automation, Manufacturing, Smart Grids, Nonlinear Systems, Power Systems, Robotics, Social Systems, Economic Systems and other. Of particular value to both the contributors and the readership are the short publication timeframe and the world-wide distribution and exposure which enable both a wide and rapid dissemination of research output.

The series covers the theory, applications, and perspectives on the state of the art and future developments relevant to systems and networks, decision making, control, complex processes and related areas, as embedded in the fields of interdisciplinary and applied sciences, engineering, computer science, physics, economics, social, and life sciences, as well as the paradigms and methodologies behind them.

Advisory Board

More information about this series at http://www.springer.com/series/15179

Oleg V. Inshakov · Agnessa O. Inshakova
Elena G. Popkova
Editors

Energy Sector: A Systemic Analysis of Economy, Foreign Trade and Legal Regulations

 Springer

Editors
Oleg V. Inshakov
Volgograd State University
Volgograd
Russia

Elena G. Popkova
Volgograd State Technical University
Volgograd
Russia

Agnessa O. Inshakova
Volgograd State University
Volgograd
Russia

ISSN 2367-3370   ISSN 2367-3389   (electronic)
Lecture Notes in Networks and Systems
ISBN 978-3-319-90965-3   ISBN 978-3-319-90966-0   (eBook)
https://doi.org/10.1007/978-3-319-90966-0

Library of Congress Control Number: 2018940405

Printed on acid-free paper

This Springer imprint is published by the registered company Springer International Publishing AG
part of Springer Nature
The registered company address is: Gewerbestrasse 11, 6330 Cham, Switzerland

*This book is dedicated to
Oleg V. Inshakov
(1952–2018)
with love and gratitude.*

Preface

Energy is rightly considered to be one of the strategically important sectors of the economy of our country, the normal functioning and further development of which depends on the economic growth of the Russian state and the strength of its positions in the foreign economic arena.

At present, Russia is one of the largest exporters of oil, gas and other energy resources to many countries of the world. Trade in energy resources occupies an important place in Russia's foreign economic turnover and can rightfully be called a key branch of national exports, whose contemporary development is directly dependent on the state of the legal foundations providing it.

The issues of legal regulation of the turnover of energy resources[1] are no less important on the scale of the entire international community, which is conditioned by the objective significance of the fuel and energy complex in the life activity of our planet. Strengthening the processes of globalization, active growth and inter-penetration of national economies striving for innovative development and unevenness in the world economic distribution of energy resources cause the emergence of complex dilemmas and new challenges in the legal regulation of international energy cooperation. The law is designed to regulate the foreign trade

---

[1]For the purposes of this study, taking into account its limited scope, the study of problems of legal regulation of obligations in the sphere of foreign trade turnover of energy resources and its unification will be limited to regulatory, contractual and local regulation of export–import activities of oil and gas companies, as well as judicial practice accompanying this activity. The concept of "energy resources" will be interpreted mainly in the narrow sense in relation to the legislative approach, as stipulated, for example, in paragraph 1 of Art. 2 of the Federal Law of 23.11.2009 No. 261-FZ "On energy conservation and on improving energy efficiency and on introducing amendments to certain legislative acts of the Russian Federation". Thus, as an energy resource as a "carrier of energy, the energy of which is used or can be used in the implementation of economic and other activities, as well as the type of energy (atomic, thermal, electrical, electromagnetic energy or another type of energy)" in this book, and also according to the majority of the normative, theoretical and empirical bases of the research used in it, it will mainly mean oil and gas as the most frequently used objects of civil rights in the field of energy resources export.

turnover of energy resources and the disagreements that arise in the course of its implementation between countries.

The objective significance, the share of the Russian fuel and energy sector in the total volume of cross-border turnover of energy resources and the dynamic development of international cooperation in the energy sector cause the relevance of the research appeal to the problems of improving legal regulation, capable of ensuring a balance of interests of all participants in these social relations—economic entities of producing countries, suppliers, consumers and transiters of energy resources.

The necessity and importance of law in this sphere are determined by the subject of regulation itself—energy as a special object of civil rights and a product that is in high demand. In foreign trade turnover, energy resources are a transnational product, which implies the complication of emerging legal relations by a foreign element and the participation of another legal system. This makes it necessary to study the issues of applicable law, that inevitably arise before the settlement of conflict and have a significant impact on the content of a foreign trade contract. In the book, researchers solve the task of conducting a comprehensive analysis of unified international and national level of substantive law, intra-state conflict regulation of foreign economic activity, international trade and judicial practice in the energy sector.

The relevance of modern scientific research on the strategy, methods and tools for the formation of adequate legal conditions for foreign trade in energy resources, including within the framework of basic civil contractual designs, can't be overemphasized. Relations arising between participants in the process of international movement of national energy resources need to be negotiated. Therefore, a significant place in the book is devoted to the study of the current legal regulation, the main directions of its development, as well as the principles and mechanisms for unifying the foreign trade contractual relations of the Russian Federation in the sphere of international turnover of national energy resources.

According to the Energy Strategy of Russia for the period until 2030 (Energy Strategy of Russia for the period up to 2030, 2009), the main vectors for the development of the fuel and energy complex are the transition to a path of energy-efficient and innovative development, as well as the integration of Russia into the world energy system. The guidelines for energy development, set by the country's main strategic act, allow us to conclude that the study of the problems of the legal regulation of the foreign trade turnover of energy resources and its unification acquires particular urgency and practical significance. Provided that the research takes into account the need to enhance high-tech, innovative activities in the energy sector and the expediency of close cooperation between the Russian Federation and international organizations and integration associations.

Modern world economic development is impossible without the implementation of innovative activities consisting in the introduction of the latest high-performance technologies and equipment, the use of advanced world experience, improving the environmentally friendly production of raw materials and the level of its processing, the transition to the use of modern raw materials, fuel, energy, based on the use of renewable and alternative energy sources. In this regard, the study reflected the issues of legal regulation of innovation in the energy sector.

Today, the external economic regulation of the energy sector can't be limited only by the domestic methods of a single country. The processes of globalization of the world economy, internationalization and liberalization of the energy sector, as well as the mutual integration of the fuel and energy complex of various states rely on close interstate cooperation, including, in the field of lawmaking of international organizations and integration associations. For international entities involved in the process of forming a balanced external economic regulation, there are many tasks, aimed at solving problems related to the fragmentary and unsystematized nature of their activities, the issues of their legal personality and competence, the lack of a unified strategy and the difficulties in developing uniform standards because of the predominance of interests a single state taken over international interests.

Unification and harmonization of the fundamental principles and norms of international law and Russian legislation regulating cooperation in the world energy market is included in the Consolidated Plan ("road map") of the state energy policy activities for the period until 2030, which ensure the implementation of the Energy Strategy of Russia.

Solutions of practical problems proposed in the book are aimed at overcoming by legal methods and means the consequences of energy crises, the destruction of energy systems, the uneven distribution of energy resources and the "energy hunger" of certain regions, lagging behind the technological development and integration of the energy sector. The identified problems are, among other things, the consequence of the prevailing disunity of the national foreign economic regulation with international norms and standards.

At the present stage, one can't fail to take into account the growing influence of economic and legal integration, acting as the starting point for the necessary legal unification of all levels of legal regulation in analysing the prospects for the development of legislation in the field of energy. In this regard, it is obvious the need to study international integration forms of cooperation and the legal foundations of foreign trade energy relations that are developed as a result of their activities. From the point of view of the development of national law and the economy of the Russian state, research in this direction is primarily of interest from international integration associations with the participation of the Russian Federation, such as the CIS, the EAES and the BRICS. In addition, it is important to understand the priority directions of development, the system, principles and essence of the legal foundations of international integration associations—the largest foreign economic partners of Russia, of which the EU is a bright representative.

The growth of the degree of energy interdependence, the strengthening of the internationalization and globalization of world energy, and the indisputability of comprehensive international cooperation to ensure energy security that is becoming widespread lead to further development and improvement of the institutions of the world energy policy of legal support and its unification in the sphere of foreign trade turnover of energy resources at the global level. The study of the legal regulation of the foreign trade turnover of energy resources in the Russian Federation, which needs to be improved and unified, is systemic in the book. All elements of the legal regulation of foreign trade in the energy sector, both general and special legislation, including unified norms of international legal regulation, strategic and framework acts, national legislation, as well as soft law, recommendatory and technical standards have been studied.

The research reflects theoretical and practical aspects of self-motivation of energy saving by business entities within their corporate market responsibility, as well as regulatory mechanisms for motivating energy conservation in the interests of ensuring sustainable development of the economy. Issues of progressive development and unification of legal regulation in the energy sector within the framework of the interstate associations are studied in the book considering the special integration potential of the energy industry of law.

Analysis of the effectiveness of existing models of the general legal energy policy of many countries, legal methods and tools used to implement it and create a unified energy regulation including in the sphere of trade turnover is an important research task, which is of great practical importance for the socioeconomic modernization of the Russian Federation.

Volgograd, Russia

Oleg V. Inshakov
Agnessa O. Inshakova
Elena G. Popkova

Contents

Contributors

Lyudmila Y. Bogachkova Department of Applied Informatics and Mathematical Methods in Economics, Institute of Management and Regional Economics, Volgograd State University, Volgograd, Russia

Aleksei V. Bogoviz Federal State Budgetary Scientific Institution "Federal Research Center of Agrarian Economy and Social Development of Rural Areas – All Russian Research Institute of Agricultural Economics", Moscow, Russia

Evgenia E. Frolova Institute of State and Law, Russian Academy of Sciences, Moscow, Russia

Alexander I. Goncharov Department of Civil and International Private Law, Institute of Law, Volgograd State University, Volgograd, Russia

Oleg V. Inshakov Department of Economic Theory, World and Regional Economics, Science of the Russian Federation, Volgograd State University, Volgograd, Russia

Agnessa O. Inshakova Department of Civil and International Private Law, Institute of Law, Volgograd State University, Moscow, Russia

Elena I. Inshakova Department of Economic Theory, World and Regional Economics, Institute of Economics and Finance, Volgograd State University, Volgograd, Russia

Igor P. Marchukov Department of Civil and International Private Law, Volgograd State University, Volgograd, Russia

Elena G. Popkova Department of World Economics and Economic Theory, Volgograd State Technical University, Volgograd, Russia

Part I
Foreign Trade Activities and Turnover of Energy Resources: Concepts, Legal Foundations and Contracting Frameworks

Importance and Sources of Legal Regulation of Foreign Trade Activities and Turnover of Energy Resources

Agnessa O. Inshakova and Igor P. Marchukov

1 Annotation

The chapter of the book substantiates the significance of the civil legal basis for the foreign trade turnover of energy resources for the social and economic modernization of the Russian state. Sources containing norms and regulations regulating this sphere of international management, as well as drafts of normative and legal documents under development and recommended for adoption by representatives of the domestic scientific doctrine are being studied. The system of legal regulation of foreign trade turnover of energy resources, as objects of civil rights, is considered by the authors from the point of view of the totality of elements, both general and special domestic legislation, norms of technical regulation, as well as international legal regulation. The basis of this is the unified substantive rules of international treaties, the customs of international business turnover and the rule of soft law, developed by various international organizations. The updated provisions of the Civil Code of the Russian Federation on the procedure for carrying out foreign trade activities by economic entities are analyzed. The doctrinal classifications of legislative acts that constitute an array of foreign trade legal regulation in the energy sector are being studied. It is noted that the international contractual practice in the Russian Federation in the field of foreign trade turnover of energy resources has developed on the basis of the existing extensive material and conflict of laws, which includes not only civil law legislation, but also the norms of administrative, export,

A. O. Inshakova (✉)
Department of Civil and International Private Law, Institute of Law,
Volgograd State University, Volgograd, Russia
e-mail: gimchp@volsu.ru; ainshakova@list.ru

I. P. Marchukov
Department of Civil and International Private Law, Volgograd State University,
Volgograd, Russia
e-mail: gimchp@volsu.ru

currency, customs and tax legislation. The expanding influence of international organizations on the development of an array of national civil-law regulation in the sphere of trade in energy resources has been established.

Are investigated in comparison with applicable in the Civil Code of the Russian Federation legislative approaches to the issues of choice of applicable law, used in international regulations: Regulation—Rome I and the United Nations Convention on Contracts for the International Sale of Goods 1980 (Vienna Convention). Based on the analysis of various theoretical approaches, as well as the practice of international commercial arbitration, the authors consider the issue on the limits of application of the provisions of the Vienna Convention to the foreign trade contract for the supply of energy resources, practical recommendations are given on the content of the contract with regard to establishing an appropriate way to exclude the application of the provisions of the Vienna Convention to contractual relations under study.

2 Materials

The basis of the analysis in this chapter of the analysis was made by sources included in the system of Russian law in the sphere of foreign trade in energy resources: program documents, normative legal acts of all levels of positive regulation, as well as model contracts and other soft law documents that have a recommendatory character. In particular, these are the main strategic program acts, first of all, the Energy Strategy of Russia for the period until 2030. The basis for the analysis of domestic regulation was the provisions of the Constitution of the Russian Federation, and Part 3 of Sect. 6 "Private International Law" of the Civil Code of the Russian Federation. In addition, the provisions of the federal laws in the sphere of energy regulating relations in the fuel and energy complex regardless of the specific industry or covering relationships in several energy sectors at once. These laws include: Federal Law of 21.07.1997 No. 116-FZ "On Industrial Safety of Dangerous Production Facilities"; Federal Law of 21.07.2011 No. 256-FZ "On the Safety of the Fuel and Energy Complex"; Federal Law of 03.12.2011 No. 382-FZ "On the state information system of the fuel and energy complex"; Federal Law of 23.11.2009 No. 261-FZ "On Energy Saving and on Improving Energy Efficiency and on Amending Certain Legislative Acts of the Russian Federation"; Law of the Russian Federation of 21.02.1992 No. 2395-1 "On Subsoil"; Federal Law of 30.11.1995 No. 187-FZ "On the Continental Shelf of the Russian Federation"; Federal Law of 21.07.2005 No. 115-FZ "On Concession Agreements".

The provisions of the Federal Laws regulating activities in a particular energy sector are explored, in particular: Federal Law of 26.03.2003 No. 35-FZ "On Electric Power Industry"; Federal Law of 27.07.2010 No. 190-FZ "On Heat Supply"; Federal Law of 21.11.1995 No. 170-FZ "On the Use of Atomic Energy"; Federal Law of 31.03.1999 No. 69-FZ "On gas supply in the Russian Federation".

The provisions of some normative acts under development, in particular, the draft Federal Law "On the peculiarities of the turnover of oil and oil products in the Russian Federation" are analyzed.

The main regulatory legal acts governing foreign economic activity in the export of energy resources include the Federal Law of 08.12.2003 No. 164-FZ "On the Basics of State Regulation of Foreign Trade Activity", Federal Law of 18.07.1999 No. 183-FZ "On Export Control"; Federal Law of 10.12.2003 No. 173-FZ "On Currency Regulation and Currency Control".

In addition to federal laws, a large number of sub-legislative acts have been analyzed among sources of Russian energy law: Decree of the Government of the Russian Federation of 27.12.2010 No. 1172 "On approval of the Rules for the Wholesale Electricity and Capacity Market and on Amending Certain Acts of the Government of the Russian Federation Regarding Regulation"; Decree of the Government of the Russian Federation of 29.12.2011 No. 1178 "On pricing in the field of regulated prices (tariffs) in the electric power industry", etc.

Among the universal, regional and bilateral international legal instruments analyzed in the work: the Vienna Convention on the International Sale of Goods of 1980; The Paris Convention for the Protection of Industrial Property of 1883, the New York Convention on the Recognition and Enforcement of Foreign Arbitral Awards of 1958; Convention on the Physical Protection of Nuclear Material; New York Convention on the Limitation Period in the International Sale of Goods of 1974; The Hague Convention on the Law Applicable to Contracts for the International Sale of Goods, 1986; Regulation No. 593/2008 of the European Parliament and the Council of the European Union "On the law subject to application to treaty obligations ("Rome I")"; European Energy Charter; Agreement to the Energy Charter; Protocol to the Energy Charter for Energy Efficiency and Related Environmental Aspects of 1994; The Mountain Charter of the Member States of the Commonwealth of Independent States of 1997; Agreement on cooperation in the field of exploration, exploration and use of mineral resources in 1997; Agreement on cross-border cooperation in the field of exploration, development and protection of the bowels of 2001, etc. The analysis of the level of international regulation is presented in the study on the basis of "soft" unified rules of recommendatory nature, such as the Rules for the interpretation of trade terms "Incoterms"; Principles of international commercial contracts Unidroit; Terms of contracts for the construction of industrial facilities, the unification of which is carried out by the International Federation of Consulting Engineers, etc.

Investigations of sources of legal regulation of the foreign trade turnover of energy resources are devoted to the works of leading Russian scientists who compiled the theoretical basis of this chapter of the book. Among them are works published in monographs, periodicals recommended by the Higher Attestation Commission of the Ministry of Education and Science of the Russian Federation, as well as those included in international analytical citation bases. Evaluation of the current conflict regulation and conflict principles, which are the basis for determining the law applicable to obligations, including in the field of foreign trade regulation of energy resources turnover is presented in the works of Asoskov (2012).

Classifications of internal statutory and legal acts are devoted to the works of Romanova (2014). The issues related to the definition of the limits of application of the Vienna Convention are considered in the proceedings: Alekseev (2014), Sukhanov (2008), Kochetov (2013). Advantages of application in the sphere of legal regulation of foreign trade turnover of energy resources of international business practices, in particular, the Unidroit Principles, are studied on the basis of works by Nikolyukin (2013), Ariffin and Yaakub (2017). Questions about the limits of the application of the norms of the Vienna Convention were also examined on the basis of the materials of international jurisprudence, in particular, the decisions of the International Commercial Arbitration Court at the Chamber of Commerce and Industry of the Russian Federation.

3   Methods

In the process of research, the general scientific dialectical method was used, and also methods are used: formal-logical, system-structural, comparison method. In the process of interpreting the results of the research, methods of synthesis, classification and generalization were used. The research also used private-science methods: formal legal, the principle of evaluating legal processes, the method of comparative analysis, and others.

4   Introduction

Energy resources are a global commodity of the modern foreign economic market. In the foreign trade turnover of the Russian Federation, the sale of energy resources and carriers is not simply an essential, but rather a predominant place. It can be safely asserted that the priority development of the fuel and energy sector as the basis of modern society in the 21st century has the character of one of the global trends in the development of the world economy and is singled out as a key branch of the national economy of Russia, as well as an important element of the country's economy.

In fact, the energy sector, ensuring the vital activity of all branches of national economy, contributes to the unification of not only the subjects of the federation, but even the strengthening of interstate socio-economic relations, and to a large extent determines the formation of the country's fundamental macroeconomic indicators.

Natural fuel and energy resources are the national property of the Russian Federation, the effective use of which forms the necessary prerequisites for the transition of the state's economic system to a path of sustainable development that ensures an increase in the level of the well-being of the population (Dudikov 2014).

Russia is one of the main energy suppliers for many foreign countries of the world. Thus, according to the Analytical Center under the Government of the Russian Federation, Russia produces about 10% of the world's primary energy, about half of this figure is exported (Energy Bulletin 2017). Prime Minister Dmitry Medvedev on December 22, 2016 before the meeting to discuss the draft "Energy Strategy of Russia until 2035" reported that the fuel and energy complex provides more than a quarter of Russia's gross domestic product and half the budget revenue (Medvedev 2017).

The successful development of the foreign trade turnover in the energy sector of Russia largely depends on the proper legal regulation of relations in the energy sector—the branch of the national economy, encompassing energy resources, the search for, extraction, generation, transformation, transmission, distribution and use of various types of energy. The necessity and importance of legal regulation in this sphere are determined by the subject of regulation—energy as an important commodity that is in high demand not only in the Russian Federation, but all over the world. In the process of creating an effective legislative framework regulating energy relations, all factors affecting the development of modern reality, as well as existing economic realities and the main trends that transform them, should be taken into account.

At the same time, it should be noted that the objective significance, the share of the Russian fuel and energy sector in the total volume of cross-border turnover of energy resources, and the dynamic development of the energy industry in Russia necessitate a continuous improvement of the legal regulation that ensures the balance of interests of all participants in public relations in the energy sector. Speaking about the system of legal regulation of energy resources as an object of foreign trade turnover, it should be noted that it includes elements of both general and special legislation, norms of technical regulation, as well as international legal regulation based on unified substantive rules of international treaties, customs of international business turnover and soft law norms, developed by various international organizations (Carr 2013).

5 Inside the State Level as an Element of the System of Legal Regulation of Foreign Trade Turnover of Energy Resources

At the domestic level, since the adoption of the Constitution of the Russian Federation at a national referendum on 12.12.1993 (The Constitution of the Russian Federation 1993), the legal regulation of relations in the sphere of foreign trade in energy resources has always been and is being given considerable attention.

The constitutional basis of Russia's energy law is the provisions of the Constitution of the Russian Federation on the support of competition, freedom of economic activity, guaranteeing the unity of the economic space, free movement of goods, services and financial resources (Article 8), free use, disposal and disposal of land and other natural resources by the owners (Article 36), on the RF federal energy systems being in charge of the RF (Article 71), on the joint jurisdiction of the RF and the RF subjects with land, subsoil legislation, environmental protection environment and issues of ownership, use and disposal of land, mineral resources and other natural resources (Article 72) (Gorodov 2012).

The development of the above constitutional provisions was followed by the adoption of a number of laws aimed at regulating various relations in the fuel and energy sector of the country. Relations in the field of foreign trade between economic entities in Russia are regulated, first of all, by the Civil Code of the Russian Federation.

In the Civil Code of the Russian Federation, in addition to general provisions on the treaty, there are special provisions governing energy relations. The norms of § 6 of Chapter 30 of the Civil Code of the Russian Federation are devoted to the contract of energy supply, the object of which is energy as a special commodity with specific properties, namely, impossibility, unlike other things (goods), accumulation and storage; continuity of the production, transmission and consumption process; impossibility of energy return after transmission; the need for special devices for its detection in the network, etc. The energy supply contract is the main contractual construction that mediates the turnover of energy goods in the Russian Federation (Balzhirov 2012).

The national legislator refers to the energy goods electric, thermal energy, gas, oil, oil products (Article 539, 548 of the Civil Code of the Russian Federation). By its legal nature, the foreign trade deal on transferring energy resources to the other party is a supply contract, with a number of features. Despite the variety of contractual relations for the transfer of energy carriers (a continuous supply contract, a supply agreement through an affiliated network, a separate power supply agreement, etc.), their legal regulation is determined by the norms of Chapter 30 of the Civil Code of the Russian Federation as having a single legal nature of purchase and sale.

In addition, part 3 of Sect. 6 "Private International Law" of the Civil Code of the Russian Federation regulates general provisions on foreign economic activity, in particular, these are issues related to:

- definition of the right to be applied to civil-law relations with the participation of foreign persons or to civil-law relations complicated by another foreign element;
- qualification of legal concepts in determining the law to be applied;
- the application of the law of the counterparty country and the establishment of its content;
- definition of the right to be applied in determining the legal status of legal entities;
- State participation in civil law relations, complicated by a foreign element;
- definition of the right to be applied to property and personal non-property relations (Marchukov 2012).

Thus, Sect. 6 of the Civil Code of the Russian Federation, substantially updated in 2013, establishes the main provisions on the procedure for foreign trade activities by business entities (Inshakova et al. 2017).

As notes A.V. Asoskov, the level of our conflict regulation in general can be compared with the leading countries. When working on amendments in 2013, all the latest trends are taken into account, in particular, some conflict solutions are borrowed from the new Regulations that appeared in the EU, primarily "Rome I" and "Rome II" (Bagaev 2016).

Equally important role in the legal regulation of the international sale and purchase of energy resources is played by the norms of public law, in particular export, currency, customs and tax legislation (Ugrin and Yanishevskaya 2016).

Among the codified laws, the Land Code of the Russian Federation (The Land Code of the RF 2001), the Urban Development Code of the Russian Federation (Urban Development Code of the RF 2005), the Tax Code of the Russian Federation (The Tax Code of the RF 1998), the Code of the Russian Federation on Administrative Offenses (The Code on Administrative Offenses of the RF 2001), and the Criminal Code of the Russian Federation (The Criminal Code of the RF 1996) play a significant role in the complex legal foundations of the foreign trade turnover of energy resources.

In the Russian Federation, many special regulatory and legal acts regulate the sphere of public activity in question with an international character.

Romanova V.V. offers a huge array of legislative acts divided into conditional groups: (1) federal laws in the field of energy that regulate certain relationships in the fuel and energy sector regardless of the specific industry (electric power industry, oil industry, gas industry, etc.) or covering relations in several energy sectors, and (2) federal laws governing relations in a particular energy sector (electricity, heat, gas, etc.) (Romanova 2014).

If we rely on the proposed dichotomy, then the first type of special federal laws in the energy sector should include, in particular, the Federal Law of 21.07.1997 No 116-FZ "On industrial safety of hazardous production facilities" (Federal Law 1997); Federal Law No 256-FZ of July 21, 2011 "On the Safety of the Fuel and Energy Complex" (Federal Law 2011b); Federal Law No 382-FZ dated 03.12.2011 "On the State Information System of the Fuel and Energy Complex" (Federal Law 2011c), Federal Law No 261-FZ of 23.11.2009 "On Energy Saving and Improving Energy Efficiency and on Amending Certain Legislative Acts of the Russian Federation" (Federal Law 2009), Law of the Russian Federation of February 21, 1992 No 2395-1 "On Subsoil" (Federal Law 1992), Federal Law of November 30, 1995 No 187-FZ "On the Continental Shelf of the Russian Federation" (Federal Law 1995b), Federal Law of July 21, 2005 No 115-FZ "On Concession Agreements" (Federal Law 2005).

The second group of federal laws that regulate activities in a particular energy sector include, in particular: Federal Law No 35-FZ of March 26, 2003 "On Electric Power Industry" (Federal Law 2003b), establishing the legal basis for economic relations in the electric power industry, the basic rights and obligations of electric power industry entities operating in the sphere of electricity consumers, the procedure for exercising state supervision in this area of economic activity; Federal Law No 190-FZ of July 27, 2010, "On Heat Supply" (Federal Law 2010), which defines the legal basis for the relations existing in connection with the transfer of heat energy, the functioning of heat supply systems, and also establishes the rights and obligations of consumers of thermal energy, heat supply organizations (Gorodov 2012), heating network organizations; Federal Law No 170-FZ of November 21, 1995 "On the Use of Atomic Energy" (Federal Law 1995c), which establishes the legal basis and principles for regulating relations arising from the use of atomic energy; Federal Law of March 31, 1999 No. 69-FZ "On gas supply in the Russian Federation" (Federal Law 1999b), which defines the legal, economic and organizational basis for relations in the field of gas supply in the Russian Federation and is aimed at meeting the state's needs for a strategic form of energy resources.

In order to prevent violations of antitrust laws in the sphere of trade in oil and petroleum products, the Federal Antimonopoly Service of the Russian Federation, with reference to the provisions of the Energy Strategy of Russia for the period until 2030, developed a draft Federal Law "On the Specifics of the Turnover of Oil and Oil Products in the Russian Federation" (Energy strategy of Russia for the period up to 2030 2009), whose purpose is to determine the basis for state regulation of trade in oil and oil products in the Russian Federation.

Activities on the export of energy resources, refers to foreign economic activity. One of the main regulatory legal acts in this area is the Federal Law No 164-FZ of 08.12.2003 "On the Basics of State Regulation of Foreign Trade Activity" (Federal Law 2009), determining the basis of state regulation of foreign trade activities, the powers of the Russian Federation and its subjects in the sphere of export and import of goods and services. Regarding the foreign trade in energy resources, the law includes provisions on the legal status of Russian and foreign persons as participants in foreign trade activities, the conclusion of international trade agreements and other RF contracts in the field of foreign economic relations, customs and tariff and non-tariff regulation, issues of licensing in the field of foreign trade in goods and the exclusive right to export and (or) import certain types of goods, to provide favorable conditions for the access of Russian persons to foreign markets (Pozdnyakova 2014).

Among the most significant normative and legal acts, we should also mention Federal Law No 183-FZ of 18.07.1999 "On Export Control" (Federal Law 1999a), which establishes the basic directions of the legal regulation of the export control system in the Russian Federation (the principles of the implementation of state policy, the legal basis for the activities of public authorities, as well as the rights, duties, responsibility of participants in foreign economic activity).

The provisions of this law are the basis for practical measures by the Government of the Russian Federation to improve the organization of oil exports.

Export operations related to the international purchase and sale of energy resources, as a rule, are carried out using foreign exchange. In this regard, it is necessary to note the existing rules of currency regulation and exchange control, which are directly related to the export of energy resources (Marchukov 2016).

Federal Law No 173-FZ of 10.12.2003 "On Currency Regulation and Currency Control" (Federal Law 2003a) provides for monitoring compliance with the currency legislation of the Russian Federation, as well as completeness and relia-bility of accounting and reporting on foreign exchange transactions.

The requirements of the currency legislation of the states to which the parties to the foreign economic contract belong, may imperatively prescribe in which cur-rency the calculations are to be made. There are also situations when one of its parties insists on fulfilling the obligation in a certain currency. The most accessible and effective tool in these cases, which makes it possible to reduce economic losses from the change in the currency of the payment from the moment of the occurrence of the obligation to the onset of its execution,—is the right of the parties to the contract to provide for the calculation and for the expression of the amount of debt different currencies. The convertibility of many foreign currencies allows a par-ticipant in international economic turnover to choose as the currency of debt that currency whose exchange rate change is minimal or most predictable (Bublik 2015).

In paragraph 2 of Article 317 of the Civil Code of the Russian Federation, the possibility of expressing a monetary obligation in foreign currency is fixed. The contract may provide for the payment of the contract in rubles in an amount equal to the amount in foreign currency. Thus, the use of foreign currency solely as a currency of debt depends on the discretion of the parties to the contract. However, the currency of payment can be foreign currency only in the cases, in the manner and under the conditions established by law or in accordance with the procedure established by it (clause 3 of Article 317 of the Civil Code of the Russian Federation).

Among the federal laws regulating activities in the energy sector, it should be noted: Federal Law of 17.08.1995 No 147-FZ "On Natural Monopolies" (Federal Law 1995a), which applies to relations arising in the commodity markets of the Russian Federation, in which subjects of natural monopolies take part. This law regulates the activities of natural monopolies, including in the sphere of trans-portation of oil, oil products, gas through pipelines; in the provision of services for the transmission of electrical energy and heat.

It should also be noted that Federal Law of 26.07.2006 No 135-FZ "On Protection of Competition" (Federal Law 2006), which covers relations related to the protection of competition in order to ensure the unity of the economic space, free movement of goods, freedom of economic activity in the Russian Federation, protection of competition and creation of conditions for effective functioning of commodity markets, and in which Russian legal entities and foreign legal entities, organizations, government bodies, as well as Federal Law No 223-FZ of 18.07.2011 "On Procurement of Goods, Works, Services by Individual Types of Legal Entities" (Federal Law 2011a), which establishes general principles for the procurement of goods, works, services and basic requirements for the purchase of goods, works, services by state corporations, state companies, public companies, natural monopoly entities, organizations that carry out regulated activities in the field of electricity supply, gas supply, and heat supply by subsidiary economic companies.

In the foreign trade of any goods, the obligatory stage is the movement of goods across the border of the state for the purpose of supply or transit. Relations in this area are regulated by customs legislation. According to Article 2 of the Federal Law of 27.11.2010 No 311-FZ "On Customs Regulation" (Federal Law 2010a, b), the customs regulation in the Russian Federation in accordance with the customs legislation of the Customs Union and the legislation of the Russian Federation is to establish the procedure and rules for regulating the customs business in the Russian Federation. The principle of freedom of transit through the territory of Russia is the main principle of customs regulation of foreign trade.

In addition to federal laws, a large number of by-laws are among the sources of energy law: Decree of the Government of the Russian Federation No. 1172 of 27.12.2010 "On Approval of the Rules for the Wholesale Electricity and Capacity Market and on Amending Certain Acts of the Government of the Russian Federation on Organization Issues" (Government of the Russian Federation 2010), Resolution of the Government of the Russian Federation No. 1178 of 29.12.2011 "On pricing in the field of regulated prices (tariffs) in the electric power industry" (Government of the Russian Federation 2011), etc.

At present, the national energy policy, which determines the sustainable development of the energy sector in the long term, acquires special significance. It embodies the goals and objectives of the development of the national energy sector, priorities and guidelines, as well as mechanisms of the state energy policy at certain stages of its implementation the energy strategy of Russia for the period up to 2030. Analysis of the provisions of this program allows us to conclude that the improvement of the regulatory and legal framework in the energy sector will follow the path of further legislation that ensures the stability, completeness and consistency of the regulatory and legal framework of this vital sphere of society.

6 The International Contractual Level of Regulation as an Element of the System of the Law of Foreign Trade Turnover of Energy Resources

International legal regulation in the field of foreign trade turnover of energy resources is mainly international agreements.

In 1991, the European Energy Charter (Kolosov and Krivchikova 1997), was approved, which outlined the main ways and principles of modern international cooperation in the field of energy (Voloshin 2015), and in 1994 the Treaty on the Energy Charter (Order of the Government of the Russian Federation 2009) was adopted—a unique multilateral international treaty containing trade, investment and transit provisions. The geographical scope of the treaty, extending to the countries of Europe, Asia (Japan, Mongolia, Turkey), the former USSR and Australia, although it takes him beyond the regional framework, but does not give him the sign of "universal recognition" and universal character.

The creation of a competitive and open market for energy products, equipment, materials and services is one of the main ways to achieve the goal of the states that signed the European Energy Charter to improve the level of security and minimize the environmental problem; removal of obstacles in energy trade, related equipment, technologies and energy-related services; providing access to development on a commercial basis and exploration of energy resources, access to local and international markets, etc.

The international instruments related to the energy sector also include the Protocol to the Energy Charter on Energy Efficiency and Related Environmental Aspects of 1994 (Paragraph 2017), designed to create a single energy space; The Mountain Charter of the Member States of the Commonwealth of Independent States, 1997 (The Mountain Charter of the Member States of the Commonwealth of Independent States 1999); Agreement on cooperation in the field of exploration, exploration and use of mineral resources in 1997 (Agreement On Cooperation in the Field of Exploration, Exploration and Use of Mineral Resources 1997); Agreement on cross-border cooperation in the field of exploration, development and protection of the bowels of 2001 (Agreement 2001), etc.

Among multilateral international treaties, the provisions of which are somehow connected with the foreign trade turnover of energy resources, first of all, the provisions of the Vienna Convention on the International Sale of Goods of 1980 (Provisions of the Vienna Convention 1990), The Paris Convention for the Protection of Industrial Property of 1883 (Paris Convention 1968), the New York Convention on the Recognition and Enforcement of Foreign Arbitral Awards of 1958 (New York Convention 1960), the Convention on the Physical Protection of Nuclear Material (Convention 1987), the New York Convention on the Limitation Period in the International Sale of Goods of 1974 (New York Convention 1974).

Unification of conflict rules was promoted by the development in 1986 of the new Hague Convention on the law applicable to contracts for the international sale of goods (Rosenberg 1996).

The Convention contains rules for determining the applicable law to contracts for the sale of goods, the scope of the applicable law, the possibility of applying the law with which the treaty, in all circumstances, has a closer link, etc. Until now, the 1986 Hague Convention has not entered into force because of existing discrepancies, not only in doctrine, but also in the practice of applying the continental and Anglo-Saxon systems of private international law, which have become an obstacle to the recognition of the Convention by a sufficient number of states. At the same time Shestakova M.P., noting the positive role of this act, correctly points out that "many formulated provisions contributed to the convergence of the positions of various states in developing common approaches to further work on the unification of conflict rules at the international legal level, and subsequently reflected in the national legislation of a number of countries." (Borisov and Vlasova 2014).

On December 17, 2009, it began to operate with the exception of certain provisions of Regulation 593/2008 of the European Parliament and of the Council of the European Union "On the law subject to application to treaty obligations ("Rome I") (MSUA 2017), reflecting contemporary requirements of international economic cooperation. The basis of the Regulations is the Rome Convention (Convention on Law, applicable to contractual obligations 2017) and the practice of its application.

Regulation "Rome I", as well as the Rome Convention, establishes the principle of autonomy of the will of the parties in choosing the applicable law (Article 3); "The contract is governed by the law chosen by the parties. The choice must be expressly expressed or definitely follows from the provisions of the treaty or from the circumstances of the case. Through this choice, the parties may indicate the applicable law for their treaty as a whole or only for a part of it."

Also Regulation "Rome I" provides provisions on the right to be applied in the absence of a choice of parties (Article 4), determining the conflict binding for the most common types of contracts used in the foreign economic sphere—purchase and sale, provision of services, rental of real estate (Kieninger 2015). For a contract of sale is the law of the country, where the seller has his usual place of residence. In those cases when the contract concluded by the parties is not included in the list of obligations that have differentiated conflict bindings or refers to mixed contracts, the rule provided for in paragraph 2 of the same article applies: the applicable law recognizes the right of the country where the usual place of residence of the party is located, which is to carry out the execution, which is of decisive importance for the treaty. There are possible exceptions to these rules, for example, for cases where the contract has clearly more close ties to another country (paragraphs 3 and 4) (Borisov and Vlasova 2014).

Thus, in the Regulations "Rome I", as in the new art. 1211 of the Civil Code of the Russian Federation, the conflict of laws rules and the principle of decisive enforcement are essential for determining the applicable law, which directly indicate the right of which state is subject to application to a contractual obligation with a foreign element, eliminating the uncertainty and indistinctness inherent in the law of closest connection.

Foreign economic activity is one of the few where a serious unification was carried out, not only conflict, but also material norms. This is explained by the desire of states to create a sufficiently effective regulator of international relations.

The UN Convention on Contracts for the International Sale of Goods, signed in 1980 in Vienna, occupies a special place among the above-mentioned fundamental international sources that regulate foreign economic transactions through the unification of substantive rules and the ones most often used to regulate relations in foreign economic transactions.

As of 2017, more than 80 countries are parties to the Vienna Convention, including the Russian Federation, acting as the legal successor to the USSR, which ratified the Convention on May 23, 1990. In accordance with Article 1 of the Vienna Convention, it applies to contracts for the sale of goods between parties whose commercial enterprises are located in different states.

It is interesting that the application of the Convention is possible not only when the party to the contract—the commercial enterprise is in the State Party (the Contracting State) but also when the law of the Contracting State is applicable under the rules of private international law (Provisions of the Vienna Convention 1990). Thus, the question of the applicability of the Vienna Convention is directly related to the actual location of the enterprise. However, this rule does not apply if the presence of parties to trade relations in different countries can't be established from a contract, business relationship or the exchange of information between them.

Article 6 of the Convention grants the parties the opportunity to exclude its application in the regulation of public relations arising from the said treaty. This rule of law, due to the lack of the necessary concretization, has led to the emergence of a number of debatable issues related to the definition of the limits of application of the Vienna Convention. Thus, at the present time in the scientific literature and in the practice of international judicial bodies, the question remains as to which way to exclude the application of the Convention to the contract of international sale is appropriate.

According to the first point of view, in order to exclude the application of the provisions of the Vienna Convention to treaty relations, it is sufficient for the parties to determine in the text of the treaty the national law of the country subject to application to these contractual legal relations.

In the scientific literature, this approach is followed by the majority of authors, including S.S. Alekseev and E.A. Sukhanov, according to which, in this case, the literal meaning of the words and expressions contained in the terms of the treaty should be accepted (Kochetov 2013).

Supporters of another scientific position, in particular G.V. Kochetov, believe that the choice of the parties as the law to be applied, the domestic law of the country, can not be considered as a refusal to apply the Convention (Kochetov 2013). In their opinion, the only appropriate way to exclude the provisions of the Convention from the normative legal acts governing the procedure for concluding a contract of sale is the clearly expressed intention of the parties set out in the text of the treaty.

The second point of view seems to us to be the most substantiated and consistent with the provisions of Article 6 of the Convention due to the following circumstances. The national legislation of the Russian Federation (Article 7 of the Civil Code of the Russian Federation) and Article 7 of the Convention establish the priority of international legal norms over the norms of national legislation, which regulates relations arising from the contract of international sale only in a part not regulated by the Vienna Convention. At the same time, the international act does not prioritize its norms over the civil legislation of the Russian Federation, depending on whether the norms of national legislation to be applied are determined on the basis of an agreement of the parties or on the basis of conflicting norms (Kochetov 2013). Thus, the issue of excluding the application of the Convention to contractual relations should be resolved by expressing the parties' intention in the text of the contract. Under such conditions, the indication in the document of the country whose law will be applied for the regulation of legal relations can't be considered as a refusal to use the provisions of the Convention.

7 Case Study

The question of the limits of the application of the norms of the Vienna Convention remains open in international judicial practice. Thus, the International Commercial Arbitration Court at the Chamber of Commerce and Industry of the Russian Federation (hereinafter referred to as the ICAC of the Russian Federation), resolving the dispute between commercial enterprises located on the territory of Ukraine and the Czech Republic, that the parties have chosen the norms of substantive law of the Russian Federation to apply to the contract of international sale, but they have not exercised the right to exclude the application of the Vienna Convention, since the text of the treaty does not contain the respective intention of the parties (Decision of the International Commercial Arbitration Court 2017). In considering a similar dispute between commercial enterprises in the UK and the Russian Federation, the ICAC at the RF CCI came to a completely opposite conclusion, that the parties realized the right granted to them by Article 7 of the Convention by including in the text of the agreement an item on the application of the norms of Russian national legislation (Decision of the International Commercial Arbitration Court 2017).

A study of judicial practice allows one to conclude that there is no uniformity in it, due to the ambiguity of the rule set forth in article 7 of the Vienna Convention. It should also be noted that the Vienna Convention does not apply to the contract for the supply of electricity through an affiliated network.

The norms of this Convention, which is typical for the majority of existing international treaties that ensure the unification of civil and commercial law, regulate only relations related to the international turnover of energy resources and do not affect operations carried out within the country. The significance of the Convention for the sphere under consideration is already characterized by the fact that it has a large number of participants (as of January 8, 2017-85 participants), among which the states are Russia's main trading partners for export deliveries of oil and gas.

The Convention, which takes into account the principles and institutions of the legal systems of various states and contributes to the creation of a single legal regime for contracts for the international sale of goods, facilitates the unification of national laws governing the international sale and purchase of goods. The Vienna Convention contains a uniform regulation of the conclusion and use of contracts for the international sale of goods, which allows the elimination of discrepancies in national legislation; promotes the elimination of inequitable discriminatory relations in international trade; promotes an unambiguous understanding by the parties of their rights and obligations; establishes a list of the objects of the contract of sale, to which the Convention does not apply; determines the basic obligations of the buyer and the seller under the contract (Garagulia 2011).

Among the researchers, there is a position on the need to create a new international document of a universal nature with a view to improving the legal foundations of world energy trade. The parties to such a legally binding agreement may be all the major producer countries, consumers and transiters of energy resources. The provisions of the Agreement should correspond to the trends of development and globalization of the world economy, internationalization and liberalization of the energy sector.

We believe that the obligatory characteristics of the international legal instrument recommended for development should be: a universal character; complexity of regulation; non-discrimination of participation; the effectiveness of the overall implementation mechanism; the absence of collisions with the norms of other international documents in this field. The agreement should cover all aspects of global energy cooperation and define: basic concepts, objectives and principles of international energy cooperation, provisions on the coordination of energy policy, trade in energy resources, competition in international energy markets, investments in the energy sector, taxation, innovation, ecology, energy security, dispute resolution, liability, etc.

In addition, the Agreement recommends defining the specifics and essential/mandatory terms of the contract for the international sale and purchase (supply) of energy resources, which primarily concern long-term contracts that are prevalent for the sphere of foreign trade turnover, for the extraction and supply of oil and gas with permanent foreign economic partners and will be discussed in detail below.

In the provision of foreign trade in energy resources, an important role is played by bilateral international agreements (Federal Law 2015). These Agreements determine the areas of cooperation for the companies participating in these Agreements, which cover, among other things, production, extraction, processing, transportation of energy resources, supply of materials, provision of services, construction of energy facilities.

It is necessary to note the growing influence of international organizations on the development of the array of national legal regulation in the sphere of trade in energy resources. For example, Russia's accession to the World Trade Organization (Federal Law 2012) leads to the need to bring national legislation regulating trade activities in general and trade in energy raw materials in particular in line with the principles and rules of the WTO.

8 Unified Customs of International Business Turnover as an Element of the Legal System in the Sphere of Foreign Trade in Energy Resources

A special place in the regulation of relations in the sphere of international trade in energy resources is occupied by the unified customs of international business turnover. The traditional elements of the foreign trade custom are, first, the duration of existence; second, the sustainability of its application; thirdly, its recognition by the states and wide popularity, otherwise the interested party will have to prove its existence (Nikolyukin 2013).

The level of international regulation is also presented by "soft" unified rules of recommendatory nature, such as the Rules for the interpretation of trade terms "Incoterms" (Vilkova 2010), Principles of international commercial contracts UNIDROIT (Komarov 2013), Terms of contracts for the construction of industrial facilities, the unification of which is carried out by the International Federation of Consulting Engineers (hereinafter referred to as FIDIC) (The official website of the International Federation of Consulting Engineers 2017) and others.

Widely distributed in international trade practice are the so-called standard contracts—these are exemplary written contracts or a set of unified conditions formulated in advance in the light of trade practices or customs accepted by the contracting parties after they have been agreed with the requirements of the particular transaction (Shitthoff 1968).

Such contracts are applicable only to certain goods or certain types of trade. They are used most often in trading activities between partners who conduct regular foreign trade operations on a long-term basis.

In the process of preparing and modeling a foreign trade transaction, only specific information changes in the existing pro forma of trade contracts. Among the main variables are the description of the product, its qualitative and quantitative characteristics, terms and method of payment, as well as transport delivery conditions, including the charter of the vessel (Golubchik and Katyuha 2016).

During the period of FIDIC's activity, various pro forma contracts were issued, actively used by the parties in the course of construction of energy facilities. For example, "The terms of the contract for the construction of objects of civil construction" ("Red" book); short form of contract ("Green" book); terms of the contract for projects "turnkey" ("Silver" book). The legal nature of the terms of contracts issued by FIDIC is of an advisory nature, the parties can follow them, or they can use other forms (Romanova 2016).

The International Institute for the Unification of Private Law in Rome developed the Principles of International Commercial Contracts (hereinafter referred to as the Unidroit Principles) containing, as stated in the preamble, "general rules for international commercial contracts".

Nikolyukin S.V. notes the following advantages of the application of the Unidroit Principles in the regulation of international commercial contracts: reflecting in them the concepts of most legal systems; the establishment of rules capable of supplementing new approaches with national legal systems, without actually requiring them to change; use of legal and technical solutions that have found application, for example, in the 1980 Vienna Convention on Contracts for the International Sale of Goods, and also provisions on those issues that were not reflected in it, i.e. replenishment of the Vienna Convention; granting to subjects of a legal relationship within the limits of the international commercial turn of an opportunity of independent definition of application to their contract of Principles of Unidroit, with or without an indication of the relevant reservation, and in this regard their application will depend on their credibility and credibility (Nikolyukin 2014).

In order to prevent misunderstandings in cross-border commercial turnover, the International Chamber of Commerce has developed a unified international rules for the interpretation of terms (INCOTERMS) (Nikolyukin 2014).

The main purpose of INCOTERMS is to systematize and standardize the terms of delivery of goods under the contract of sale. INCOTERMS is one of the most important international documents of unofficial codification of international trade customs, where unified rules on the interpretation of trade terms widely used in world trade are formulated (Anufrieva 2013).

INCOTERMS identifies some types of international sales contracts based on the distribution of rights and obligations of trade partners for the transport of goods, the implementation of rules relating to the export and import of goods, its transit through third countries, the transition of the risk of accidental loss or damage to the goods during transportation from the seller to the buyer.

Particular importance in the process of concluding and executing a foreign trade contract for the supply of energy resources in the Russian Federation are the customs of merchant shipping, the possibility of using them is enshrined in Article 285 of the RF Merchant Shipping Code. Direct transfer of energy resources to the other side, first of all, oil and oil products, coal is carried out by sea vessels of the tanker type, in this regard, the application of marine trade customs becomes particularly relevant.

If the parties choose Russian law as applicable to the contract, the parties will be able to refer to the York-Antwerp Rules of 1994 (Comite Marine International Antwerpen 2017), which regulate the procedure for compensation of losses in the event of a general maritime accident involving merchant vessels. Attention should be paid to the fact that the reference in the foreign trade contract to well-known and accepted trade terms, indicates the application to the contract of trade custom, denoted by these terms. Parties have the right to also include in the contract other instructions regarding the application of trade customs.

The norms of technical regulation play an important role for effective legal regulation of relations in the energy sector. These include regulations aimed at ensuring environmental and other security, establishing technical and technological conditions, standards and systems for assessing the compliance of the Eurasian Economic Union and the Customs Union. These provisions are developed in accordance with the WTO Agreement on Technical Barriers to Trade (The Committee on the Integration of Trade and Customs Policy and the WTO 2017).

9 Conclusion

In conclusion, it should be noted that currently the existing regulatory legal acts of energy legislation represent a stable system of legal support for the country's fuel and energy complex. In the sphere of foreign trade turnover of energy resources, this system consists of elements of general and special legislation, norms of technical regulation, unified substantive rules of international legal contracts, international business practices and a law of recommendation developed by various international organizations.

Contractual practice in the Russian Federation in the field of foreign trade turnover in the energy sector has taken into account the extensive material and conflict legal framework, including not only civil and legal legislation, but also the norms of administrative, export, currency, customs and tax legislation.

Based on the analysis of various theoretical approaches to the limits of the application of the norms of the Vienna Convention on Contracts for the International Sale of Goods of 1980 concerning the proper way to exclude the application of the provisions of the Convention to the contract for the international sale of energy resources, as well as the practice of international commercial arbitration concluded that it is not enough simply to define in the text of the treaty national law applicable to these contractual legal relations.

As the only proper way to exclude the provisions of the Convention from the normative legal acts governing the procedure for concluding a contract of sale, it is recommended to reflect in its text the clearly expressed intention of the parties.

If this condition is agreed, the following wording should be used: "The norms of the legislation of the Russian Federation are applicable to the relations of the parties under the contract. The parties exclude the application of the United Nations Convention on Contracts for the International Sale of Goods of 1980 to the legal relations that govern the contract."

The certainty and uniqueness of the wording of the treaty will make it possible to exclude possible potential contradictions in the partnership relations of the parties to a foreign trade transaction, and also to avoid differences in the understanding of the formulations that make up the contents of the treaty and prevent a correct understanding of the true will of its parties. In addition, such an approach will help to effectively protect the rights and legitimate interests of foreign trade participants, reduce the risks of unfair conduct by one of the parties to the treaty, and in the event of a dispute and in the process of resolving it, will limit the discretion of the court in interpreting its terms. The proposed approach does not contradict the current Russian legislation and is consistent with the following regulatory requirements: Article 1, paragraph 1 (a), Article 7 of the Vienna Convention, Part 4 of Article 15 of the Constitution of the Russian Federation, Article 7 of the Civil Code of the Russian Federation.

References

Agreement. (2001). *On Cross-border cooperation in the field of exploration, development and protection of mineral resources, (approved by the Decree of the Government of the Russian Federation of May 23, 2001 no. 405).* (Russian).

Agreement between the Government of the Russian Federation and the Government of the Slovak Republic on cooperation in the field of long-term oil supplies from the Russian Federation to the Slovak Republic and transit of Russian oil through the territory of the Slovak Republic. (2014). Information Legal Portal Garant. Retrieved January 21, 2017, from http://base.garant.ru/70779144/. (Russian).

Agreement on Cooperation in the Field of Exploration, Exploration and Use of Mineral Resources. (1997). Information Legal Portal Garant. Retrieved January 21, 2017, from http://base.garant. ru/1148100/. (Russian).

Alekseev, S. S. (2014). *Civil law in questions and answers*. Moscow: Publishing House Prospect.

Anufrieva, L. P. (2013). *International public law: A textbook for Bachelors: for students of higher education institutions studying in specialty 021100 Jurisprudence*. Moscow: Prospekt.

Arifin, A., & Yaakub, N. I. (2017). Principles of the international commercial contract for Unidroit, as rules of law governing cross-border contracts. *Advanced scientific letters, 23*(1), 478–481.

Asoskov, A. V. (2012). The basics of conflict law. Moscow: Infopropic Media.

Bagaev, V. A. (2016). The level of our conflict regulation can be compared with the leading countries. *Law, 8,* 6–16.

Balzhirov, B. V. (2012). On the use of the energy supply contract in the energy sector of Russia. *Civil Law, 3,* 45–48.

Borisov, V. N., & Vlasova, N. V. (2014). *Certain types of obligations in private international law: Monograph*. Moscow: INFRA-M.

Bublik, V. A. (2015). Currency regulation in Russia: Present and future. *Russian Law Journal, 6,* 170–179.

Carr, I. (2013). *International trade law* (5th ed.). London: Rutledge.

Code of the Russian Federation. (2001). *On Administrative Offenses, (approved of the Federation Council on December 26, 2001 no. 1)*. (Russian).

Comite Marine International Antwerpen. (2017). *York-Antwerp Rules 1974. Reglesd 'Yorketd' Anvers 1974*. Retrieved January 1, 2017, from http://docs.cntd.ru/document/1900880.

Convention. (1987). *On the Physical Protection of Nuclear Material (approved of the Decree of the Presidium of the Supreme Soviet of the USSR of May 4, 1983 no. 9236-X)*.

Convention on Law, Applicable to Contractual Obligations. (2017). *Electronic fund of legal and normative technical documentation*. Retrieved September 4, 2017, from http://docs.cntd.ru/document/901889343.

Decision of the International Commercial Arbitration Court at the Chamber of Commerce and Industry of the Russian Federation of 30.01.2015 in case No. 100/2014 2016, Information Legal Portal Garant. Retrieved January 21, 2017, from http://base.garant.ru/71259966/. (Russian).

Decision of the International Commercial Arbitration Court at the Chamber of Commerce and Industry of the Russian Federation of 28.04.2009 in case no. 120/2009 2017. Information Legal Portal Garant. Retrieved January 21, 2017, from http://base.garant.ru/5214111/. (Russian).

Dudikov, M. V. (2014). Legal regulation of the accounting of production and consumption of fuel and energy resources in the Russian Federation. *Environmental Law, 3,* 33–36.

Energy Bulletin. (2017). Development of oil transportation. Retrieved January 21, 2017, from http://www.energyland.info/files/library/062016/6ab618bbc5fd23f5b6afe5531bc52d4d.pdf.

Energy strategy of Russia for the period up to 2030 2009, *(approved by the order of the Government of the Russian Federation on November 13, 2009 no 1715-p)*. (Russian).

Federal Law. (1995a). *On Natural Monopolies (adopted by the State Duma on July 19, 1995 no. 147-FL)*. (Russian).

Federal Law. (1995b). *On the Continental Shelf of the Russian Federation, (adopted by the State Duma on November 30, 1995 no. 261-FL)*. (Russian).

Federal Law. (1995c). *On the Use of Atomic Energy, (adopted by the State Duma on October 19, 1995 no. 170-FL)*. (Russian).

Federal Law. (1997). *On industrial safety of hazardous production facilities, (adopted by the State Duma on June 20, 1997 no. 116-FL)*. (Russian).

Federal Law. (1999a). *On Export Control, (approved by the Federation Council on July 2, 1993 no. 183-FL)*. (Russian).

Federal Law. (1999b). *On Gas Supply in the Russian Federation, (approved by the Federation Council on March 31, 1999 no. 311-FL)*. (Russian).

Federal Law. (2003a). *On Currency Regulation and Currency Control, (adopted by the State Duma on December 30, 1995 no. 173-FL).* (Russian).

Federal Law. (2003b). *On Electric Power Industry, (approved by the Federation Council on March 26, 2003 no. 35-FL).* (Russian).

Federal Law. (2003c). *On the Basics of State Regulation of Foreign Trade Activity, (approved by the Federation Council on November 26, 2003 no. 164-FL).* (Russian).

Federal Law. (2005). *On Concession Agreements, (approved by the Federation Council on July 15, 2005 no. 115-FL).* (Russian).

Federal Law. (2006). *On Protection of Competition, (approved by the Federation Council on July 26, 2006 no. 135-FL).* (Russian).

Federal Law. (2009). *On Energy Saving and on Improving Energy Efficiency and on Amending Certain Legislative Acts of the Russian Federation, (adopted by the State Duma on November 23, 2009 no. 261-FL).* (Russian).

Federal Law. (2010a). *On Customs Regulation, (approved by the Federation Council on November 27, 2010 no. 311-FL).* (Russian).

Federal Law. (2010b). *On Heat Supply, (approved by the Federation Council on July 27, 2010 no. 190-FL). (Russian).*

Federal Law. (2011a). *On Procurement of Goods, Works, Services by Individual Types of Legal Entities, (approved by the Federation Council on July 11, 2011 no. 223-FL).* (Russian).

Federal Law. (2011b). *On the Safety of the Fuel and Energy Complex, (approved by the Federation Council on July 13, 2011 no. 256-FL).* (Russian).

Federal Law. (2011c). *On the state information system of the fuel and energy complex, (approved by the Federation Council on November 29, 2011 no. 382-FL).* (Russian).

Federal Law. (2012). *On Ratification of the Protocol of Accession of the Russian Federation to the Marrakesh Agreement Establishing the World Trade Organization of April 15, 1994, (approved by the Federation Council on July 18, 2012 no. 126-FL).* (Russian).

Federal Law. (2015). *On Ratification of the Agreement between the Government of the Russian Federation and the Government of the People's Republic of China on Cooperation in the Sphere of Natural Gas Supplies from the Russian Federation to the People's Republic of China on the Eastern Route, (approved by the Federation Council on April 29, 2015 no. 106-FL).* (Russian).

Federal Portal of Draft Normative Legal Acts. (2017). *As of January 2017 is not included in the State Duma of the Federal Assembly of the Russian Federation.* Retrieved January 21, 2017, from http://regulation.gov.ru. (Russian).

Garagulya, M. I. (2011). The problem of unification of the norms regulating foreign economic contractual relations. *Lawyer, 10,* 27–31.

Golubchik, A. M., & Katyuha, P. B. (2016). Next specific features of contract work in oil trading. *Russian Foreign Economic Journal, 6,* 18–28.

Gorodov, O. A. (2012). *Introduction to energy law: A training manual.* Moscow: Prospekt.

Government of the Russian Federation. (2010). *On Approval of the Rules for the Wholesale Electricity and Capacity Market and on Amending Certain Acts of the Government of the Russian Federation on Organization Issues, (approved by the Decree of the Government of the Russian Federation of December 27, 2010 no. 1172).* (Russian).

Government of the Russian Federation. (2011). *On pricing in the field of regulated prices (tariffs) in the electric power industry 2012, (approved by the Decree of the Government of the Russian Federation of December 21, 2011 no. 1178).* (Russian).

Inshakova, A. O., Goncharov, A. I., Mineev, O. A., & Sevostyanov, M. V. (2017). Amendments to the civil code of the russian federation: contradictions of theory and practice, *contributions to economics. Russia and the European Union: Development and Perspectives, 8,* 147–153.

Killinger, E. M. (2015). Assignment of claims in the Rome statute. *ERA Forum, 16*(2), 181–196.

Kochetov, G.V. (2013). Concerning the application of decisions in the sphere of commercial activities. *Bulletin of Volgograd State University. Series 5. Jurisprudence, 3*(20), pp. 136–139.

Kolosov, Y. M., & Krivchikova, E. S. (1997). *Current international law. T.3,* Publishing house Moscow Independent Institute of International Law, Moscow.

Komarov, A. S. (2013). *Principles of international treaties of UNIDROIT 2010 treaties*, Statute, Moscow.

Law of the Russian Federation No. 2395–1. (1992). *On Subsoil, (approved of the Rossiyskaya Gazeta of May 5, 1992, no. 102).* (Russian).

Marchukov, I. P. (2016). Legal basis for Russia's foreign trade in energy as a priority export industry. *Bulletin of Volgograd State University. Series 5. Jurisprudence, 1*(30), 137–144.

Medvedev, D. A. (2017). *The fuel and energy sector accounts for more than a quarter of Russia's GDP and half of budget revenues,* The official website of the TASS news agency. Retrieved January 21, 2017, from http://tass.ru/ekonomika/3897636. (Russian).

MSUA. (2017). *Regulation No. 593/2008 of the European Parliament and of the Council of the European Union On the law subject to application to treaty obligations (Rome I).* Retrieved April 26, 2017, from http://eulaw.edu.ru/documents/legislation/collision/dogovornoe.htm#_ftn1.

New York Convention. (1960). *On the Recognition and Enforcement of Foreign Arbitral Awards, (approved of the Decree of the Presidium of the Supreme Soviet of the USSR of August 10, 1960).* (Russian).

New York Convention on the Limitation Period in the International Sale of Goods. (1974). CIGS Russia. Retrieved January 21, 2017, from http://www.cisg.ru/dok/limit_ru.pdf (Russian).

Nikolyukin, S. V. (2013). Systematization and standardization of international trade operations (for example, INCOTERMS-2010). *Lawyer, 6,* 12–17.

Nikolyukin, S. V. (2014). Unification of the rules of international sale and purchase (on the example of the Unidroit Principles). *Jurist, 4,* 36–41.

Order of the Government of the Russian Federation. (2009). *On sending notice of the intention of the Russian Federation not to become a party to the Energy Charter Treaty, as well as the Energy Charter Protocol on Energy Efficiency and Related Environmental Aspects of the System, (approved by the Order of the Government of the Russian Federation of July 30, 2009 no. 1055-r).* (Russian).

Paragraph. (2017). *Protocol to the Energy Charter on Energy Efficiency and Related Environmental Aspects of 1994.* Retrieved January 21, 2017, from http://online.zakon.kz/Document/?doc_id=1027144#pos=0;0.

Paris Convention. (1968). *For the Protection of Industrial Property 1883. (approved of the Decree of the Presidium of the Supreme Soviet of the USSR of 19 September, 1968 no. 3104-VII).* (Russian).

Pozdnyakova, L. M. (2014). *Legal regulation of foreign economic activity (Russian civil and international private law)*: Uch. pos., Norma, Moscow.

Provisions of the Vienna Convention. (1990). *On the International Sale of Goods of 1980, (approved of the Decree of the Supreme Soviet of the USSR of May 23, 1990 no. 1511–1).* (Russian).

Resolution of the Government of the Russian Federation. (1994). *On the signing of the Energy Charter Treaty and related documents, (approved by the Decree of the Government of the Russian Federation of December 16, 1994 no. 1390).* (Russian).

Romanova, V. V. (2014) Energy Law. A common part. *The special part,* Lawyer, Moscow.

Romanova, V. V. (2016). Peculiarities of legal regulation of foreign economic transactions in the energy sector. *International Public and Private Law, 1,* 16–21.

Rosenberg, M. G. (1996). Contract of international sale and purchase. Modern practice of imprisonment. Settlement of disputes, International Center for Financial and Economic Development, Moscow.

Shithoff, K. (1968). Combining or harmonizing the law with standard contracts and general conditions. *International and Comparative Law Quarterly, 17,* 551–770.

Sukhanov, E. A. (2008). Civil law. In *Obligations* (In 11 volumes, vol. 3). Textbook, Wolters Kluwer, Moscow.

The Committee on the Integration of Trade and Customs Policy and the WTO. (2017). *The World Trade Organization Agreement on Technical Barriers to Trade (consolidated text).* Retrieved January 22, 2017, from http://rgwto.com/wto.asp?id=3668&doc_id=2099. (Russian).

The Constitution of the Russian Federation. (1993). *(was adopted by popular vote on December 12, 1993).* (Russian).

The Criminal Code of the Russian Federation. (1996). *(the Code was put into effect on June 13, 1996, no. 63-FZ).* (Russian).

The Land Code of the Russian Federation. (2001*). (approved of the Collection of legislation of the Russian Federation of October 29, 2001 no. 44, art. 4147.).* (Russian).

The Mountain Charter of the Member States of the Commonwealth of Independent States. (1999). *(approved of the Bulletin of international treaties of March 27, 1999 no. 7, pp. 3–8).*

The official website of the International Federation of Consulting Engineers 2017. Retrieved January 22, 2017, from www.fidic.org. (Russian).

The Tax Code of the Russian Federation. (1998). *(approved by the Federation Council on July 17, 1998).*

Ugrin, T. S., & Yanishevskaya, Y. A. (2016). 'Foreign trade contract: Form, customs privileges, education. *The Science Scientific Staff, 1,* 67–70.

Urban Development Code of the Russian Federation No. 190-FZ. (2005). *(approved of the Collection of legislation of the Russian Federation of 3 January, 2005 no. 1(1), art. 16).* (Russian).

Vilkova, N. G. (2010). *International rules for the use of trade terms Incoterms 2010 Publication ICC No. 715,* Infotropic Media, Moscow.

Voloshin, V. I. (2015). *The EU-Russia Energy Dialogue,* The Russian-European Center for Economic Policy (RECEP). Retrieved November 22, 2015, from http://www.recep.ru/files/publ/trends2.pdf. (Russian).

Basic Concepts of Legal Regulation of Foreign Trade Activities of Energy Resources Turnover

Agnessa O. Inshakova and Igor P. Marchukov

1 Annotation

The chapter defines the concepts that are significant for the legal regulation of foreign trade in the energy sector and the correct interpretation of the content and conclusions of the monographic research, as well as for the development of civil law, for science, and for branch of legislation. These include the concepts: "energy legislation", "object of foreign economic transactions", "foreign trade activities in the energy sector", "foreign trade turnover of energy resources", "export control", as well as "energy resource in relation to foreign trade turnover".

Authors offer their own original definitions of foreign trade activities in the energy and foreign trade turnover of energy resources, correlating among themselves as a whole and part. The formulated concepts will contribute to the building up of civil-law theoretical knowledge, the elimination of possible difficulties in the process of the formation of uniform law-enforcement practice, which are associated with the lack of legislative consolidation of these key concepts for the legal regulation of the research area of management concepts, as well as the uncertainty of doctrinal approaches on this issue.

It is pointed out that it is possible to consolidate the formulated concepts in the proposed by the researchers to the development of an integrated interdisciplinary codified act—the Energy Code of the Russian Federation.

A. O. Inshakova (✉)
Department of Civil and International Private Law, Institute of Law,
Volgograd State University, Volgograd, Russia
e-mail: gimchp@volsu.ru; ainshakova@list.ru

I. P. Marchukov
Department of Civil and International Private Law, Volgograd State University,
Volgograd, Russia
e-mail: gimchp@volsu.ru

© Springer International Publishing AG, part of Springer Nature 2019
O. V. Inshakov et al. (eds.), *Energy Sector: A Systemic Analysis of Economy,*
Foreign Trade and Legal Regulations, Lecture Notes in Networks and Systems 44,
https://doi.org/10.1007/978-3-319-90966-0_2

2 Materials

The legislative basis for the study was drafted by the Federal Law "On the Specifics of the Turnover of Oil and Petroleum Products in the Russian Federation," which introduces into the regulatory field a number of new terms analyzed in the research. And also a number of existing normative legal acts containing some concepts that directly or indirectly relate to the sphere of foreign trade turnover of energy resources. Also served as a basis for the development and further specification of definitions of the main concepts and categories, formulated by the authors and of general importance for this area of legal regulation. For example, Federal Law No. 164-FZ "On the Basics of State Regulation of Foreign Trade Activity".

The energy legislation of the Russian Federation as a set of normative legal acts and an integrated branch of law regulating activities in the field of energy has become the subject of research in the work of Popondopulo (2008), Yakovlev and Lakhno (2011), Melekhin and Dudikov (2012), Gorodov (2012) and others.

Definition and characteristic features of the concept of energy objects as objects of foreign economic transactions are investigated in the works of Romanova (2014) Theoretical approaches to the question of determining the legal nature of energy and energy resources as objects of civil rights were studied using the example of scientific works Braginsky (1967), Korneev (1956), Tarkhov (1979), Sukhanov (1998) and others.

3 Methods

In the course of the research, both general scientific methods (dialectical method of cognition, analysis, synthesis, analogy, formal-logical method, etc.) and private-science methods (comparative-legal, historical-legal, systemic, formal-legal, functional, analytical, and etc.).

4 Introduction

In the dynamic development of Russia's energy industry, as well as the high proportion of the Russian fuel and energy sector in the overall volume of cross-border turnover of energy resources, one of the primary tasks facing the Russian state is the continuous development and improvement of the regulatory and legal regulation of public relations in the energy sector, proceeding from modern requirements of a civil turn.

It is practically impossible to solve this problem without resorting to legal science, studying the basic legislative and doctrinal approaches to the formation of the investigated area of legal regulation and defining the basic concepts for it.

So, back in the 20 s of the last century, the national energy law was being drawn up on a doctrinal level. Energy as a key branch of the national economy and social relations arising within the framework of this branch are the basis of the set of norms of its constituent subjects. The fact that in 2017, in the list of ciphers of specialties and passports approved by the Higher Attestation Commission of the Ministry of Education and Science of the Russian Federation, a new passport of the specialty with the code number 12.00.07 and a formula containing an indication of corporate and energy law (The official website of the Higher Attestation Commission of the Ministry of Education and Science of the Russian Federation 2017).

The general concepts of legal regulation of foreign trade turnover of energy resources, the definition of which is necessary both for the purposes of its development and improvement, and for the purpose of a correct interpretation of the content and conclusions of this study, are the concepts of "energy legislation", "object of foreign economic transactions", "foreign trade in energy", foreign trade turnover of energy resources", "export control", as well as "energy resource in relation to foreign trade turnover."

5 Conditionality of the Complex Nature of Energy Legislation

Availability of a wide array of sources of legal regulation of foreign trade turnover of energy resources, a brief legal review of which was presented in the previous chapter, allowed representatives of the scientific community to come to a conclusion about the complex nature of energy legislation due to the nature of regulated relations in the energy sector. Indeed, it is difficult to disagree with this.

So, Popondopulo V.F. defines energy legislation as a set of normative and legal acts regulating activities in the energy sector. Energy legislation, the scientist notes, can't be of a complex nature, due to the fact that in the energy sector there are horizontal (civil-law) and vertical (administrative-legal) relations, for example, to establish control over compliance with special requirements for energy organizations, antimonopoly, tariff and other regulations in this area (Popondopulo 2008).

A similar view is held by V.F. Yakovlev, P.G. Lakhno (Yakovlev and Lakhno 2011); E.S. Melekhin, M.V. Dudikov (Melekhin and Dudikov 2012), Gorodov (2012), Baydin (2009), considering the energy legislation of the Russian Federation as an integrated complex branch of legislation.

In the last few years there have been tangible attempts by the authorities to develop and adopt special regulations in the energy sector. An example is the already mentioned in the previous paragraph draft of the Federal Law "On the Specifics of the Turnover of Oil and Oil Products in the Russian Federation." (Federal Antimonopoly Service of Russia 2017). The project introduces a number of new terms into the standard field, including: "turnover of oil and oil products", "petroleum products", "motor fuel", "retail trade in motor fuel," "wholesale trade in petroleum products."

At the time of writing the thesis, the draft law has not yet been adopted, so it is premature to talk about the existence of a conceptual-categorical apparatus developed in legislation, even with respect to certain areas of legal regulation of energy.

However, there are a number of existing regulations that contain certain concepts that directly or indirectly relate to the sphere of foreign trade turnover of energy resources and can serve as a basis for the development and further specification of definitions of basic concepts and categories that are of general importance for this area of legal regulation.

For example, Federal Law of 08.12.2003 No. 164-FZ "On the Basics of State Regulation of Foreign Trade Activity" (Federal Law 2003) defines foreign trade in goods, which is understood as the activity for the implementation of transactions in the field of foreign trade in goods, services, information and intellectual property (paragraph 4 of article 2).

Electric energy and other types of energy in accordance with paragraph 2 of article 2 of the law are attributed to goods that are the subject of foreign trade activity.

It should be agreed with V.V. Romanova, that the object of foreign economic transactions can be energy facilities, which are understood as objects through which the extraction, processing, production, transportation, storage of energy resources are carried out (Romanova 2014).

In addition to types of energy, energy facilities and energy carriers, energy equipment and energy services (transit, construction, engineering services, etc.) can be the subject of a foreign trade transaction in the energy sector (Rusnak 2015).

Proceeding from the foregoing, and also with the aim of forming the conceptual apparatus of this study, such concepts as foreign trade activity in the field of energy and foreign trade turnover of energy resources are of particular interest.

6 The Concepts of "Foreign Trade in the Energy Sector" and "Foreign Trade Turnover of Energy Resources": The Relationship and the Need to Consolidate

The legislator does not define these concepts. Indirectly, Russian legislation mentions only such a category as "a participant in foreign economic activities carrying out the export of crude oil from the Russian Federation outside the customs territory of the Customs Union by pipeline transport" (Government of the Russian Federation 2013).

At the same time, both the Russian and foreign legislators will define the concepts of "energy resources" (Article 2 of the Federal Law of 23.11.2009 № 261-FZ "On energy conservation and on improving energy efficiency and on introducing amendments to certain legislative acts of the Russian Federation". Appendix EM to the European Energy Charter), transit of energy materials and products and international transit (Article 7 of the European Energy Charter, Federal Law of 08.12.2003 No. 164-FZ "On the Basics of State Regulation of Foreign Trade Activities"), etc.

In the legal literature there is also no definition of these concepts.

In order to fill this gap, we propose, under foreign trade activities in the energy sector, to understand the activities in the field of foreign trade in various types of energy (nuclear, thermal, electric, electromagnetic energy or other type of energy), energy objects, energy equipment, energy services, and also energy carriers, whose energy is used or can be used for international economic and other activities in the world market.

Accordingly, on the basis of the definition of foreign trade in goods, formulated in paragraph 7 of Article 2 of the Federal Law of 08.12.2003 No. 164-FZ "On the Basics of State Regulation of Foreign Trade Activity", and taking into account paragraph 2 of Article 2 of the law, it can be concluded that the foreign trade turnover of energy resources is the import and (or) export of energy resources as a special type of goods to the international market.

These definitions relate to each other as a whole and part of it. On the need to consolidate these concepts, there are two opposing positions. According to the first point of view, the definition of terms used in legal norms, strictly speaking, is not a function of the legislator, since science is called upon to solve this task (Vinichenko 2013). Other researchers defend the position on the need to legislate terms for the purpose of uniform enforcement (Tsvetkov 2007).

We adhere to the second point of view and consider it expedient to consolidate the developed concepts in the legislation.

However, these definitions can't be introduced into the current Federal Law of 08.12.2003 No. 164-FZ "On the fundamentals of state regulation of foreign trade activities" because this normative act is aimed at regulating general issues of foreign trade activity. Illogical is also the proposal to include the developed concepts in special laws regulating social relations in specific energy sectors (in the gas industry, in the field of electric power, heat supply, etc.), as well as in bills (for example, the Draft Federal Law "On the Specifics of the Turnover of Oil and Oil Products in the Russian Federation", etc.) (Federal Antimonopoly Service of Russia 2017).

We believe that it would be possible to consolidate these concepts in a complex interdisciplinary codified act, for example, the Energy Code of the Russian Federation, which is mentioned in the works of modern researchers (Bykov 2013; Zankovsky 2016; Lakhno and Yakovlev 2012; Kononov 2014).

Considering the many special regulatory and legal acts that make up the energy legislation and the urgent problem of ensuring the necessary level of its unification, which deserves support, it is necessary to recognize the idea of creating the Energy Code of the Russian Federation, designed to provide comprehensive legal regulation of relations in the energy sector and contributes to the creation of a single unified system of legal regulation in this area.

The Energy Code of the Russian Federation is essential for the further formation and improvement of unique features, conditioned by the subject of legal regulation, the regulatory and legal array that serves as the basis for the legal provision of relations in the sphere of organization and functioning of the energy sector, including in the international arena.

7 The Basic Concept-Categorical Apparatus in the Sphere of Foreign Trade Turnover of Energy Resources

The concept of "export control" is closely connected with the foreign trade turnover of energy resources, under which, according to article 1 of Federal Law of 18.07.1999 No 183-FZ "On Export Control" (Federal Law 1999), we mean a "set of measures, ensuring the implementation of the procedure established by this Federal Law, other federal laws and other normative legal acts of the Russian Federation for the implementation of foreign economic activities in relation to goods, information, works, services, the results of intellectual activity (rights to them), which can be used to create weapons of mass destruction, their means of delivery, other types of weapons and military equipment, or in the preparation and (or) perpetration of terrorist acts".

This law contains the only legal definition of the term "foreign economic activity", which should be understood as foreign trade, investment and other activities, including production cooperation, in the field of international exchange of goods, information, works, services, and the results of intellectual activity (rights to them).

The study of this topic of the paragraph is impossible without considering the question of the legal nature of energy resources or energy as objects of civil rights, which is still debatable.

To date, there are three basic concepts of energy: one of them suggests to consider energy as a kind of things; the second proceeds from the premise that energy is the property benefit of an immaterial nature; the third considers energy as an independent object of legal regulation (Lavrik and Frolov 2014; Baldwin et al. 2017).

Speaking about the correlation between the concepts of "energy resources", "energy resources" and "energy," it should be noted that in legal literature and regulations they are used as identical. Thus, paragraph 1 of Article 2 of the Federal Law of 23.11.2009, No 261-FZ "On Energy Saving and on Improving Energy Efficiency and on Amending Certain Legislative Acts of the Russian Federation" defines the energy resource, as a "carrier of energy whose energy is used or can be used for economic and other activities, as well as the type of energy (atomic, thermal, electric, electromagnetic energy or other type of energy)". Similarly, the concept of energy resource and energy is the same as the Treaty on the Energy

Charter of 1994, as in the Annex of EM "Energy Materials and Products" contains the following sections:

- nuclear energy;
- coal, natural gas, oil and oil products, electricity;
- other types of energy (Simões and Blusun 2017).

The existence of various interpretations of the concept of "energy resources" and its relationship with the notion of "natural resources" in international legal acts and norms of the domestic legislation of the Russian Federation attests, on the one hand, on the complexity and ambiguity of these concepts, and on the other—their close connection.

8 The Legal Nature of Energy and Energy Resources as Objects of Civil Circulation

The existing contractual practice of different countries, as well as practice in the field of foreign trade activity, considers energy resources as a commodity transferred under a contract of sale. In particular, in the Russian Federation, the supply of energy resources is formalized by concluding a special type of contract of sale and purchase—an energy supply contract. In accordance with this contractual practice, energy resources should be considered as things, in the sense that this legal category is assigned in Article 128 of the Civil Code of the Russian Federation.

Thus, it is necessary to answer the question of whether it is possible to include energy and energy resources in the category of things. This problem of interpretation first arose in the late XIX—early XX century, in connection with the development of gas trade. On this issue in science there are different views. So, in the opinion of M.I. Braginsky, electric, thermal energy, as well as gas (not enclosed in the storage) can be considered as a material object in energy supply contracts (Braginsky 1967). The point of view of S.M. Korneev, which consists in the fact that energy can't be a thing, since it is the ability to produce useful work, to provide the necessary conditions for entrepreneurial and any other activity (Korneev 1956). In the literature, Korneev's position was developed by V.A. Tarkhov, who believed that the goal of supplying energy was not the transfer of things, but the transfer of material goods (Tarkhov 1979). E.A. Sukhanov, in his work, on the contrary, believes that any kind of energy resources is a thing, since labor is spent on production and transfer of energy and gas, as well as other resources, they have value, are products of special industries, have a quantitative and qualitative assessment, like all other things (Sukhanov 1998). Russian legislation, although it does not contain direct rules on the attribution of energy resources to a particular object of civil rights, but indirectly relates this type of goods to the category of things. So, the current Civil Code of the Russian Federation in Article 539 refers to an energy supply contract to supply contracts, which means recognition of this product as an object of property rights, things.

From the position of civil law, energy is difficult to be considered as a material object, good, however, the legislator holds a point of view according to which energy can be identified as an object of civil law (Leal-Arcas et al. 2015).

According to D.N. Mikhalev, the term energy should be understood as two types of goods: electrical (and thermal) energy, as well as energy resources—oil, gas, water, steam and products produced on their basis (Mikhalev 2012).

Opposite point of view is followed by S.A. Tebenkova, in the opinion of which the concept of "energy resources" is absolutely identical with the concept of "energy carriers" and is different from the term "energy", since energy sources can be considered in a static situation, and the energy itself is in a dynamic state (Tebenkova 2013).

In our opinion, it is worthwhile to agree with S.A. Tebenkova about the need to differentiate these two concepts in connection with the following physical characteristics of energy:

- the inability to see or touch;
- impossibility of storage in a static state;
- limiting the possibility of possession and disposal;
- the coincidence of the moment of production and consumption in time;
- the appropriation of energy is possible through appropriate means of its transmission or belonging to sources.

9 Conclusion

We believe that in the legal qualification of energy and energy resources (energy carriers), it is necessary to differentiate these concepts.

However, all kinds of energy resources as well as energy can act in civil circulation as a commodity, subject of contractual relations and determined by the concept of a thing.

Therefore, on the basis of the foregoing and taking into account the definition of the energy resource formulated in Cl. 2 of the Federal Law of 23.11.2009 № 261-FZ "On energy conservation and on improving energy efficiency and on introducing amendments to certain legislative acts of the Russian Federation" (Federal Law 2009), the concept of energy resource in relation to foreign trade turnover should be defined as a special type of goods, which is a carrier of energy, the energy of which is used or can be used for international economic activities and other activities in the world market related to import and/or export, as well as the type of energy (atomic, thermal, electric, electromagnetic energy or other type of energy).

Thus, the foreign trade turnover of energy resources is the import and (or) export of energy resources as a special type of goods representing the energy carrier whose energy is used or can be used for international economic and other activities in the world market, as well as the type of energy (atomic, thermal, electrical, electromagnetic energy, or other form of energy).

References

Baldwin, E., & Rountree, V. et al. (2017). Distributed resources and distributed governance: Stakeholder participation in demand side management governance. *Energy Research and Social Science, 39*, 37–45.

Baydin, E. V. (2009). International trade in electricity. *The Russian Foreign Economic Bulletin, 3*, 3–5.

Braginsky, M. I. (1967). *General doctrine of economic contracts*. Minsk: Science and Technology.

Bykov, A. G. (2013). On the draft energy code of the russian federation. *Energy Law, 2*, 6–8.

Federal Antimonopoly Service of Russia. (2017). *Draft Federal Law On the Specifics of Oil and Oil Product Flow in the Russian Federation , an explanatory note to it and a draft resolution on the approval of disclosure standards*. Retrieved May 3, 2017, from http://fas.gov.ru/documents/documentdetails.html?id=606. (Russian).

Federal Law. (1999). *On Export Control, (approved by the Decree of the Government of the Russian Federation of July 18, 1999 no. 183-FL)*. (Russian).

Federal Law. (2003). *On the Basics of State Regulation of Foreign Trade Activity, (approved by the Decree of the Government of the Russian Federation of December 8, 2003 no. 164-FL)*. (Russian).

Federal Law. (2009). On Energy Saving and on Improving Energy Efficiency and on Amending Certain Legislative Acts of the Russian Federation *(approved by the Decree of the Government of the Russian Federation of November 23, 2009 no. 183-FL)*. (Russian).

Federal Portal of Projects of Normative Legal Acts. (2017). Retrieved January 21, 2017, from http://regulation.gov.ru. (Russian).

Gorodov, O. A. (2012). *Introduction to energy law: A training manual*. Moscow: Prospekt.

Government of the Russian Federation. (2013). *On the procedure for confirming the production and control of the amount of crude oil, in respect of which special formulas for calculating the rates of export customs duties on crude oil, as well as on recognizing as invalid the Resolution of the Government of the Russian Federation of July 16 2009 No. 574, (approved by the Decree of the Government of the Russian Federation of March 29, 2013 no. 277)*. (Russian).

Kononov, P. I. (2014). On the systematization of russian administrative law and legislation. *Administrative Law and Process, 6*, 6–11.

Korneev, S. M. (1956). *Electricity supply agreement*. Moscow: Gosyurizdat.

Lakhno, P. G., & Yakovlev, V. F. (2012). *The Energy Code of the Russian Federation is a fundamental legal document regulating relations in the fuel and energy sector. The rule of law: Issues of formation*, Statute, Moscow.

Lavrik, T. M., & Frolov, S. A. (2014). *Legal regulation of relations in the field of energy: A textbook for students studying in the direction 030900 Jurisprudence* , FGBOU HPE TSTU, Tambov.

Leal-Arcas, R., Alemany Ríos, J., et al. (2015). The European Union and its energy security challenges: engagement through and with networks. *Contemporary Politics, 21*(3), 273–293.

Melekhin, E. S., & Dudikov, M. V. (2012). *Energy Law*. Moscow: Logos.

Mikhalev, D. N. (2012). The legal nature of the category energy, as a kind of object of civil legal relations. *Bulletin of the Kalmyk University, 2*(14), 9–12.

Popondopulo, V. F. (2008). *Energy Law and energy legislation: General characteristics, trends of development. Energy and law*, Lawyer, Moscow.

Romanova, V. V. (2014). *Energy Law. A common part. Special part: Textbook*, Lawyer, Moscow.

Rusnak, U. (2015). European energy charter. *Economist (United Kingdom)*, no. 411(8942).

Simões, F. D., & Blusun, S. A. (2017). Legal stability and renewable energy investments. *Review of European, Comparative and International Environmental Law, 26*(3), 298–304.

Sukhanov, E. A. (1998). *Civil law T. 1.*, BEK, Moscow.

Tarkhov, V. A. (1979). *Soviet civil law. Part 2*, Publishing House: SSU, Saratov.

Tebenkova, S. A. (2013). Electrical energy as an object of civil rights. *Bulletin of the Udmurt University, 2–4*, 6–10.

The All-Russian Classification of Fixed Assets. (1994). *(approved by the Decree of the State Standard of the Russian Federation No. 359 of December 26, 1994)*. (Russian).

The official website of the Higher Attestation Commission of the Ministry of Education and Science of the Russian Federation. (2017). *Passports of scientific specialties*. Retrieved January 21, 2017, from http://vak.ed.gov.ru/316. (Russian).

Tsvetkov, S. B. (2007). On the essence and concept of civil turnover. *Freedom Right Market, 5*, 147–154.

Vinichenko, Y. V. (2013). On the notion of civil turnover (statement of the problem). *Siberian legal Bulletin, 3*, 29–35.

Yakovlev, V. F., & Lakhno, P. G. (2011). *Energy law of Russia and Germany: A comparative legal study*. Moscow: Lawyer.

Zankovsky, S. S. (2016). Energy legislation in the general system of business and legal norms. *Journal of Russian Law, 8*(236), 141–145.

Foreign Trade Contracts as a Contractual Basis for International Civil Circulation of Energy Resources: Concept, Types, Content

Agnessa O. Inshakova and Igor P. Marchukov

1 Annotation

The chapter deals with the various civil legal contractual arrangements envisaged by the current Russian legislation, which regulate the foreign trade turnover of energy resources. The central place in the work is the contract of foreign trade (international) purchase and sale (delivery) of goods, as the main type of foreign economic transactions in the investigated sphere of economic interrelations. The legal nature of the foreign trade contract for the supply of energy resources, the peculiarities of the legal regime of energy resources as an object of civil rights, features of this type of foreign economic transaction in the list of other legal constructions of civil law nature, applicable in the sphere of turnover of energy resources, primarily oil and gas. It is noted that the long-term nature of foreign trade contracts for the supply of energy resources is preferable for economic entities of the Russian Federation.

Legislative and doctrinal approaches, contractual practice in the sphere of legal relations under study in the context of the definition and agreement by the parties of foreign trade contracts of the supply of energy resources of mutual rights and obligations at its conclusion are analyzed and commented on. Essential conditions of such contracts, arising from the meaning of the Vienna Convention and Article 432 of the Civil Code of the Russian Federation, are disclosed. Based on the analysis carried out by the authors, recommendations are given on the formulation and content of a foreign trade agreement, taking into account the advantages of a

A. O. Inshakova (✉)
Department of Civil and International Private Law, Institute of Law,
Volgograd State University, Volgograd, Russia
e-mail: gimchp@volsu.ru; ainshakova@list.ru

I. P. Marchukov
Department of Civil and International Private Law, Volgograd State University,
Volgograd, Russia
e-mail: gimchp@volsu.ru

© Springer International Publishing AG, part of Springer Nature 2019
O. V. Inshakov et al. (eds.), *Energy Sector: A Systemic Analysis of Economy, Foreign Trade and Legal Regulations*, Lecture Notes in Networks and Systems 44,
https://doi.org/10.1007/978-3-319-90966-0_3

long-term nature of the obligation relations, trends in the development of foreign trade energy regulation and special properties of the goods. A list of conditions for foreign trade long-term contracts for the supply of energy resources is formulated, the coordination of which, in the authors' opinion, will facilitate maximum effective fulfillment of obligations by the parties and minimization of the risks of conflicts arising from violations of the rights and legitimate interests of the parties.

2 Materials

The basis of the regulatory framework of the study was the provisions of the Vienna Convention on the International Sale of Goods of 1980 and the Civil Code of the Russian Federation, primarily Article 432, as well as the Federal Law of 30.12.1995 No. 225-FZ "On Production Sharing Agreements", the Tax Code of the Russian Federation and others.

The results of scientific research Neshataeva (2002) contributed to the study of the essential features forming the concept of "international commercial transaction" or "cross-border transaction".

The study of foreign economic transactions as the basis for international trade, the foundation and the central link of international commodity circulation was carried out on the basis of theoretical approaches formulated in the works of Dmitrieva (2013), Inshakova (2015).

In analyzing the problems of transaction qualification as an external economic transaction in the context of regulation of international economic activity, the authors relied on the works of Kanashevsky (2008, 2009), Shestakova (2014), Doronina (2014), Makovsky and Khlestova (2012). Peculiarities of the legal regulation of foreign economic transactions in the energy sector, taken into account by the authors, are analyzed in the works of Romanova (2016). Specificity and types of legal contractual constructions of civil law character corresponding to different stages of oil turnover were studied using the example of Ivanov (2013).

Various doctrinal approaches to the definition of substantive contract terms have been studied through scientific publications Belov (2013), Braginsky and Vitryansky (1997), Ioffe (1975), Lunts (1973), Novitsky and Lunts (1950), Sergeev and Tolstoy (2000), Sukhanov (2000).

The questions of the form of the foreign economic transaction and the law applicable to it were analyzed on the basis of Article 1209 of the Civil Code of the Russian Federation and Article 162 of the Civil Code of the Russian Federation, as well as scientific works of Abova et al. (2007), Boguslavsky (2016), Rodionova et al. (2017).

Based on the works of Doronina (2014), Vlasov and Kovalenko (2015), Makovsky and Khlestova (2012), a comparative analysis of various types of agreements in the field of subsoil use, including the production sharing agreement and the concession agreement, as well as other agreements that are concluded in the process of the emergence of economic relations with respect to petroleum products of consumption.

The empirical basis of the study was a contractual practice, reflecting the specifics, as well as significant conditions prevailing for the sphere of foreign trade turnover, long-term contracts for the extraction and supply of oil and gas to the EU countries the largest Russian companies such as PJSC Gazprom, PJSC Lukoil, and PJSC NK Rosneft.

3 Methods

In the process of research, general scientific methods were used, such as: formal-logical, dialectical, system-structural, cognition, empirical description. In interpreting the results of the study, a synthesis, classification and generalization method was used. The work also used private-science methods: formal legal, the principle of evaluating legal processes, comparative legal, etc.

4 Introduction

Relations arising between different subjects of international business activities associated with the movement and trade turnover of energy resources need to be negotiated. The current Russian legislation does not contain a definition of either an external economic contract, a foreign trade contract, or a foreign economic transaction.

The foreign trade contract is the main form of registration of foreign trade transactions.

The term "foreign economic transaction" is complex and means the activity of subjects of private international law in the field of international exchange of goods, works, services aimed at establishing, changing and terminating civil rights and obligations (Fedoseyeva 2005). Under the term international commercial transaction (Neshataeva 2002) or cross-border transaction, more common in international trade practice, understand the transaction that predetermines the private entrepreneurship in the field of international trade and foreign economic activity. These transactions are concluded between parties whose commercial organizations are located on the territory of different states (Dmitrieva 2013).

Foreign economic transactions are the basis of international trade, the foundation, the central link of international commodity circulation (Inshakova 2015).

The problem of qualifying a transaction as a foreign economic one is one of the most important issues in the regulation of international economic activity. The external nature of foreign economic activity is associated with the presence of a "foreign element", the location of parties in different countries, the movement of goods, services, financial resources across the customs border of the Russian Federation and the performance of work in the territory of a foreign state (Kanashevsky 2008; Kanashevsky 2009).

The contract of foreign trade (international) purchase and sale of goods is the main type of foreign economic transactions. The number of such contracts prevails in the aggregate of world economic interrelations (Kurlychev 2008; Kanashevsky 2009; Prokushev 2012; Shestakova 2014). Active work on the unification of rules for the international sale of goods began in the mid-fifties of the last century (Asoskov 2012; Vilkova 2002; Zvekov 2004; Lunts1973; Rosenberg 2004; Rosenberg, 2006; Sadikov 1986, Marysheva 2000; Funk 2005; Makovsky and Khlestova 2012).

5   Foreign Trade Agreement on the Purchase and Sale of Energy Resources and Other Types of Foreign Economic Transactions in the Energy Sector

From the above, we can conclude: the contract for foreign trade in the sale of energy resources is a document that formalizes a foreign trade transaction, by virtue of which the seller is obliged to transfer energy resources to the buyer's property, and the buyer must accept this product and pay a certain amount of money for it. In its legal nature, a foreign economic transaction for the transfer of energy resources to the other party is a supply contract that has a number of specific features conditioned by the object of the contract (Braginsky and Vitryansky 1997; Klein 1976; Agarkov et al. 1944; Yurchenko 1961; Puginsky 2016; Sukhanov 2008). Transmission of energy resources can be carried out by using various contractual forms, depending on the needs of the buyer and seller. So there is the usual supply, the transfer of energy through the connected network, continuous supply, as well as the contract for energy supply (Seinaroyev 2000; Galkina 2014), etc.

Legal regulation of foreign economic relations is characterized by very dynamic development, both at the national and international levels. When concluding and executing foreign economic transactions, the peculiarities of the legal regime of their facilities, the specific features of the legal status of the subjects, the requirements of tax and customs regulation, which must be taken into account, are established. All listed features of foreign economic transactions have their own specifics in the energy sector.

Thus, in the energy sector, various types of foreign economic transactions are concluded. Among them: deals, the subject of which is purchase and sale, transportation, transportation of energy resources (Krasavchikov 1985), purchase and sale of power equipment; transactions, the subject of which is the design, construction of energy facilities and others. Specifics of legal regulation are conditioned, first of all, by the specific nature of the main object of foreign economic transactions—the energy resource. This specificity is reflected in the peculiarities of the legal regime of energy resources as an object of foreign economic transactions (Romanova 2016).

The key agreements that regulate the relationships between the participants in the gas and oil business in the field of industrial relations associated with the system of circulation of mined raw materials are contracts for processing, transportation and supply. These agreements ensure the movement of gas, oil and oil products from production sites to final consumption sites. In the sphere connected with the transportation of oil through the backbone grids, the following main types of agreements exist: agreements for the transportation of oil through trunk pipelines, agreements on the storage of oil in the system of trunk pipelines, agreements on the assignment of rights to transport oil.

The types and nature of agreements depend on many factors: on the type of products, on the specifics of the parties to the agreements, on the mode of delivery of energy resources, and so on (Krylov and Stolyarova 2016).

To date, in the gas and oil industry can be divided into three stages of value-added activities:

1. Upstream (upstream, from English—the upper stream)—extraction and operation;
2. Midstream (midstream, from English—medium flow)—transportation;
3. Downstream (downstream, from English—bottom stream)—processing and marketing.

In this case, it is worth making a reservation that to the stage of the downstream in the oil industry is traditionally referred to the refining of oil and the sale of oil and oil products (Zeng et al. 2017). In relation to the gas industry, processing of natural and associated petroleum gas occurs before the transportation of gas (for example, liquefaction or compression of natural gas), while oil is processed at refineries, usually after transporting it through main oil pipelines from production sites (Kozlov 2015).

For each of the stages, due to the specific nature of the emerging social relations, the application of special legal contractual constructions of a civil law character is inherent.

At the stage of "upstream" the following contractual relations are used:

(A) Contractual arrangements necessary to gain access to the search and development of oil and gas fields:

- production sharing agreement (an agreement concluded between a foreign oil company (contractor) and a state corporation for exploration and exploitation of a field within a certain contract area);
- agreement on a joint venture to market natural gas (the subject of the agreement is the establishment of a joint venture between the gas producer and the sales organization);
- a contract for a contract (for example, a drilling contract on the high seas);
- a sublease agreement (the owner of a share in a joint operating agreement or a holder of a license to drill and/or extract oil transfers his rights to the sub lease on certain conditions and in exchange for certain actions of the latter).

(B) Contractual constructions directly aimed at the exploration and development of oil and gas fields—oil service agreements, the essence of which is to ensure the activities of the oil company with various types of work and services necessary for the development of oil and gas fields.

Ivanov A.A. distinguishes the following distinctive signs of the oilfield services contracts concluded at the stage of "upstream":

(1) these contracts are entrepreneurial, which is due to the purpose of any oil company—the extraction of profits;
(2) in the contracts there are always involved specific objects of law—subsoil, hazardous waste, rights, etc. The peculiarities of the legal regime of these objects are due to various reasons, but in all cases they are united by the presence of public interest aimed at protecting the interests of society and (or) the state;
(3) a special subject matter of transactions, namely, the mandatory participation in the contractual relationship of the user of the subsoil.

With the development of the economy and the complication of contractual relations in economic turnover, special types of contractual relations appear in the field of rendering oilfield services. They are new to Russian practice, but they are often used in foreign countries. For example, agreements for large-scale capital projects (EPC (engineering, procurement, construction or elements), general contractor services (design, procurement, construction or components) and EPCM (engineering and procurement services and construction management) contracts, design services and procurement, as well as construction management (Ivanov 2013).

At the "midstream" stage, transportation contracts are used that mediate the movement of gas, oil and oil products from the production sites to the final consumption sites. In the sphere connected with the transportation of oil through the backbone grids, there are the following main types of agreements (Shevchenko 2013b): the agreement on the transportation of oil through main oil pipelines, the agreement on the storage of oil in the system of trunk pipelines, the agreement on the assignment of rights to transport oil.

At the stage of "downstream" the following contractual relations are used:

• gas purchase and sale contract;
• an agreement on the sale (to supply) of crude oil, the contract for the supply of petroleum products;
• an agreement on a joint venture for the marketing of natural gas (a framework agreement whereby a producer (supplier) of natural gas attracts a local company to represent its interests and conclude gas purchase and sale contracts with end customers. Cooperation is carried out on the basis of profit sharing) (Peace and right 2017).

The stage of selling (selling) the products of the oil and gas industry, to which the dissertation research was devoted, is no less important than exploration, development and production. The sale of gas, oil and oil products is the link connecting the producer and the consumer. These relations are the most important element of production, and if they are not settled, this can lead to short supply or supply exceeding demand, which ultimately can cause fluctuations in the global economy (Kiyashko 2007). In this connection, it is impossible to overestimate the importance of oil and gas purchase and sale contracts.

Therefore, in the context of this study and as the most widely used contractual design, first of all, the contract for the sale and purchase (supply) of energy resources will be considered.

6 The Contents of the Foreign Trade Contract for the Sale and Purchase of Energy Resources: Significant and Recommended Conditions

In the domestic legal literature different opinions are expressed regarding the list of essential conditions of the contract of sale. These include the conditions of the price, the procedure for settlements, the quality and quantity of goods, the place and time of delivery, the scope of liability, the procedure for resolving disputes. Obligatory conditions of the contract of sale and purchase will always be the item (name and quantity of goods) and price. However, this list is nowhere defined. Including the UN Convention does not specify what conditions for such contracts are significant. Article 14 states only that "a proposal is definite if it identifies the goods and directly or indirectly establishes the quantity and price or provides for the procedure for determining them." From this it follows that the Vienna Convention refers to the number of essential conditions characterizing the subject of the contract (name and quantity of goods), as well as the price condition or the method of its determination.

In the Russian legislation, Article 432 of the Civil Code of the Russian Federation stipulates that: "A contract is considered concluded if an agreement has been reached between the parties in the form required in the cases to be filed in all the essential terms of the contract. Essential are the terms of the subject matter of the contract, the conditions that are called in the law or other legal acts as essential or necessary for contracts of this type, as well as all those terms concerning which an agreement must be reached upon the request of one of the parties".

The above definition generally coincides with the views of pre-revolutionary (Sinaisky 1917; Meyer 1997) and soviet researchers (Ioffe 1975; Novitsky and Lunts 1950), as well as it does not raise questions and is not critically perceived by representatives of modern legal doctrine. There are a number of modern researchers who believe that the conditions that are necessary and sufficient for concluding a

contract without agreement, or in case of whose falling off, a treaty of the appropriate type can't be considered concluded, are recognized as essential.

At the same time, there are two positions on the issue of the purposes and main
purpose of the category "essential conditions of the contract" in the domestic
literature, which, in our opinion, are mutually complementary.

Thus, according to the first, the binding nature of certain conditions of the
contract can, in particular, guarantee the protection of the interests of the weak party
(Braginsky and Vitryansky 1997). In all cases, unless we are talking about the need
to protect the interests of one of the parties to the contract or the society as a whole,
the inclusion of a condition in the material is aimed at creating guarantees of the
certainty of the relationship between the parties, which is directly interested in civil
turnover (Braginsky et al. 1997).

Another position boils down to the fact that the main purpose of the material
conditions, based on Article 432 of the Civil Code of the Russian Federation, is to
determine the minimum, under which the contract is deemed to be concluded
(Osmolovskaya 2013).

It should be noted that in relation to energy contracts, a list of their essential
conditions is among the discussion topics in modern legal doctrine. There are
different points of view on the question. According to one of them, an essential
condition for contracts of this type is called only an object (Sergeev and Tolstoy
2000). Opposite position, according to which the essential conditions of treaties in
this area include: the subject of the contract, the conditions of quantity, quality of
energy, price, as well as the conditions for ensuring the maintenance and safe
operation of networks, appliances and equipment (Vitryansky 1999).

We believe that the consolidation in the legislation of a broadly scattered list of
essential conditions of treaties in the sphere in question is inexpedient and unrea-
sonable (Shevchenko 2013a; Osmolovskaya 2013).

Proceeding from this, according to the sense of the Vienna Convention, article
432 of the Civil Code of the Russian Federation and the legal nature of the foreign
trade contract (contract) for the purchase and sale (supply) of energy resources to its
essential conditions, the condition of the subject and price of the contract, as well as
other conditions defining its content, indicated by the parties to the contract.

However, in the context of this research, taking into account the trends in the
liberalization of energy regulation in the foreign trade sphere, the special properties
of the goods and the goal of elaborating more detailed recommendations on the
formulation and content of a long-term foreign trade contract (contract) for the
purchase and sale (supply) of energy resources in addition to the essential conditions of such a contract (contract), all the conditions that are mandatory/necessary
for the most effective fulfillment of the obligations assumed by the parties under the
contract and minimizing the risks of conflict arising from violation of the rights and
interests of the parties should be taken into consideration and included in the
content of the contract.

In the context of the research of the topic, contractual practice reflecting the
specifics, as well as significant conditions (Shevchenko 2013b), prevailing for the
sphere of foreign trade turnover, long-term contracts for the extraction and supply

of oil and gas to EU countries by Russian companies, such as PJSC Gazprom, PJSC Lukoil, PJSC NK Rosneft and other major organizations. So among essential conditions it is necessary to allocate:

1. Period of validity of the contract. As a rule, the contract begins to act immediately after its signing. However, as in any civil law contract, the parties to the contract have the right to determine the date of the beginning of deliveries, and also to agree that the beginning of the contract will be preceded by a preliminary notification (from the month) from the Supplier or the Buyer.
   Often contracts are of a long-term nature. In these contracts, there are standardized periods, which vary from 20 to 25 years. This condition significantly allows to guarantee a certain insured return on investment, especially in the gas transportation infrastructure. It also guarantees reliable supplies for the importer.
2. Specifications. Specifying the point of delivery of goods is a prerequisite for a long-term contract. For gas supply contracts, as a rule, the delivery point is a gas distribution station or a compressor station near the interstate border.
3. Scope of supply. The basis for determining the volume of delivery is the daily amount (Daily Contract Quantity—DCQ). The maximum daily quantity (Maximum Daily Quantity—MDQ) available to the consumer for selection is determined in the contract either in absolute values or as a percentage of the daily volume. Swing—the value equivalent to the ratio of the peak value of energy supply to the average. "Load Factor" is the inverse of Swing. Based on these values, the total volume of the product delivered during the year is determined.
4. Balancing of deliveries on requests. The consumer usually applies for the supply of energy "on the day ahead." You can get more of the product specified in the application, but you will have to pay a higher price for it.
5. The condition of "take-or-pay", the essence of which is the obligation to pay all or part of the goods, regardless of the volume of its real consumption. In such contracts, the determination of the amount of performance depends on the Buyer and the parties are not known in advance. In other words, take-or-pay contracts differ in that they contain the Buyer's commitment to the minimum number of "orders" (the minimum amount of receipt of performance) (Law. RU 2017). In the event that the Buyer selects the goods less than the agreed take-or-pay level, his payment is considered as a charge for balancing. In the process of execution of the contract due to shortage of goods, force majeure, the implementation of the Carry Forward mechanism, violation of the specifications of the delivered goods and the take-or-pay level may be reduced at the Seller's request. This condition allows the supplier to insure against the risk of "volume". Recently, this condition is accompanied by a number of additional amendments, for example, such as the transfer of volume from one period to another, thus ensuring greater flexibility of contract terms.
   From the point of view of economic theory, the contractual model of take-or-pay serves to reduce the risk of the supplier (contractor, contractor) from "under loading" capacities. For example, when building a gas pipeline to a new

country, a party needs guarantees that through the gas pipeline, gas will be procured in at least a certain amount, otherwise the construction itself will not be appropriate. Reduction of this risk in theory leads to lower prices for the Buyer. A contractual take-or-pay model is also necessary to obtain funding for such a project, since a bank or other funding organization needs guarantees that the company will be able to service the credit debt from future revenues (Law. RU 2017).

We believe that the take-or-pay condition is preferable for long-term foreign trade contracts in the energy sector, as it ensures a stable guaranteed receipt of revenue to the supplier (Lakhno and Zekker 2011). The ability of this condition to act as a means of long-term consolidation of the supplier's share in the sales market is directly indicated by the researchers in the list of positive results due to its consolidation in the contract (Gudkov 2008).

6. Mechanisms of flexibility. Make Up is the mechanism by which the Buyer, in the event of payment of the take-or-pay amount, has the right to select a part of the unselected quantity of goods in the future contract period. Usually its size is 100% of the take-or-pay amount, but with various additional restrictions. Carry Forward is a mechanism, in some sense, a reverse make-up, in which the Buyer, in the event of the selection of more gas than take-or-pay, the generated "surplus" can use to compensate the take-or-pay obligation of the future (not necessarily the next) contractual. Usually carry-forward is considered in shares from take-or-pay, limiting the upper limit of the mechanism for selection (this is not MDQ).

7. Systematization of under delivery. Parties to long-term contracts pay considerable attention to the reasons for short delivery, depending on the reasons, the liability is differentiated. Intentional shortage often leads to a decrease in the take-or-pay level, covering by the Supplier of all (including indirect) losses to the consumer and giving the buyer the opportunity to terminate the contract. Unintentional shortage (due to an accident or other problems) will cause a decrease in the take-or-pay level, and also lead to a possible reduction in the price of subsequent deliveries. Force majeure (natural disasters, government decisions and strikes), as a rule, lead only to a decrease in the level of take-or-pay (Del Río 2017; Gas Forum 2017).

7 Form and Structure of the Foreign Trade Contract for the Sale of Energy Resources

Important when concluding a foreign trade contract is determining the form of the foreign trade transaction. According to experts, the approaches of the national legislator, both in Russian and in Soviet private international law, have always differed conservatively in relation to the establishment of the form of a foreign economic transaction. For a long time, by virtue of clause 2 of article 1209 of the

Civil Code of the Russian Federation, the Russian law was mandatory for the form of the foreign economic transaction. Thus, this one-sided conflict rule referred to the provisions of paragraph 3 of Article 162 of the Civil Code of the Russian Federation, according to which non-compliance with the simple written form of the foreign economic transaction entailed its invalidity (Abova et al. 2007; Boguslavsky 2016). Russian jurisprudence and doctrine qualified this rule as super-operative.

However, from September 1, 2013, in accordance with Federal Law of 07.05.2013 No 100-FZ, paragraph 3 of Article 162 of the Civil Code of the Russian Federation is recognized as invalid, and from November 1, 2013, according to the Federal Law of 30.09.2013 No 260-FZ, the new edition of Article 1209 of the Civil Code of the Russian Federation came into force—the form of the transaction is subject to the law of the country where it was committed. A transaction made abroad can't be declared void due to non-compliance with the form, if the requirements of Russian law are met.

The imperative conflict principle on the application of Russian law and the mandatory written form of any foreign economic transaction involving Russian persons gave way to flexible conflict regulation, based on the principle of favor negotii ("in favor of (formal validity) of the agreement"), fixed in the new wording of paragraph 1 of Article 1209 of the Civil Code of the Russian Federation and well known in foreign law and order. The approach implemented in the course of civil law reform allows us to speak about its gradual liberalization and focus on more flexible, international standards for regulating private-law relations.

Under the new rules, if the requirements to the form of foreign economic transaction, established in at least one of the applicable law and order are observed: the law of the country where the transaction was made and the RF law, if one of the parties to the transaction is a person with a personal law of the Russian Federation or a law subject to application to the transaction itself, the transaction due to non-compliance with the form can't be recognized as invalid. The new version of clause 1 of Article 1209 of the Civil Code of the Russian Federation consolidated an alternative conflict of laws rules aimed at preserving the validity of the transaction. According to articles 1210 and 1217 of the Civil Code of the Russian Federation, the law regulating the content of the transaction can be chosen by the participants in the transaction. Consequently, the parties to the transaction, when choosing the contractual statute simultaneously affect the decision of the issue of the law applicable to the form of this transaction (Asoskov 2014).

The structure of the foreign trade contract is individual, but in accordance with the recommendations the contract must contain mandatory sections, such as the subject of the contract; price and payment procedure; delivery time; acceptance of goods; force majeure circumstances; consideration of disputes; responsibility of the parties; final provisions (Semenikhin 2015).

It should be noted that long-term contracts are the basis of stability and reliability of gas and oil supplies. Only such contracts can guarantee the importer reliable and uninterrupted supply of energy for a long period of time, and the exporter—the payback of multibillion investments for the implementation of large export projects.

8   Case Study

As an example, such contracts as contracts signed by PJSC Gazprom with the
following Italian gas companies can be cited: Premium Gas S.p.A.—for a period of
13 years (with an option for an extension of 5 years), ERG S.p.A.—for a period of
10 years (with an option for an extension of 10 years), Sinergie Italiane S.r.l.—for a
period of 10 years, EGL—for a period of 20 years. In total, under these contracts
for the period up to 2028 (without options), the Group will supply 28.6 billion cubic
meters. m of gas (Gazprom 2017a).

Rosneft and Petro Vietnam Oil Corporation (PV OIL) (a subsidiary of
Petrovietnam) in 2016 signed a long-term contract for the supply of oil. The con-
tract provides for the delivery of up to 96 million tons of oil to 2040 to the
Vietnamese side (Rosneft 2017).

To further enhance the competitiveness of Russian gas, Gazprom is improving
contract work. The company also uses alternative forms of trade—in particular, gas
auctions, which allows increasing the volume of gas sold. The meeting of the Board
of Directors stressed that the combination of long-term contracts with new forms
and mechanisms of trade would allow Gazprom to strengthen its market positions
and increase revenues from the export of Russian gas (Gazprom 2017b).

However, preference is given still to long-term contracts, to the adherence of
European consumers, in particular in relations, with PJSC Gazprom, in particular,
the extension of export contracts with such Western partners as: GDF SUEZ
(France) until 2030, E. ON Ruhrgas (Germany) until 2035, RWE Transgas (Czech
Republic) until 2035, ENI (Italy) until 2035. Contracts were concluded with WIEE
(Switzerland) until 2030, with Conef Energy (Romania) until 2030, with German
company WIEH (Germany) until 2027, with Italian Premium Gas (Italy) until 2024
(Gazprom in questions and answers 2017).

When signing a bilateral agreement, it is absolutely clear that each state seeks to
bring it closer to its own model of the document, as a result of which there can not
exist two absolutely identical contracts. However, analyzing the energy supply
agreements concluded between Russia and the EU countries, it should be noted that
the similarity of most of them is that the price of gas is the amount associated with
the cost of oil.

9   Foreign Economic Contracts for Subsoil Use: General
and Specific Characteristics of Certain Species

One of the varieties of international contracts are, so-called, oil contracts—a kind of
subsoil use contracts. Oil contracts have in common with contracts for subsoil use
the features and distinctive features that allow them to separate into a separate group
of contracts.

The subsoil use contract is an agreement between the parties on the implementation of a certain type of subsoil use (mineral exploration, mining, combined exploration and mining, and others) (Moroz 2007).

The specific features of the subsoil use contract are as follows:

(1) the state in the person of its bodies is necessarily the party to the subsoil use contract;
(2) the stage of concluding a contract is preceded by the holding of a competition of investment programs for the right of subsoil use or conducting direct negotiations;
(3) the grounds and procedure for concluding, amending, terminating or terminating a subsoil use contract shall be established by special legislation (legislation on subsoil and subsoil use), and not by civil legislation (Moroz 2007). The existence of such signs is not inherent in ordinary civil-law contracts.

The parties to the subsoil use contract are the subsoil user and the competent authority.

The peculiarity of the subsoil use contract is that the draft contract is coordinated not only by the parties, but also by various ministries and departments, and is subject to compulsory economic, tax examination and mandatory state registration in the state body that concluded the contract.

Thus, we should talk about a more stringent legal regulation, manifested in the supervision and supervision of the type of international contractual relations under consideration by state bodies representing the interests of one or another national jurisdiction.

Despite the fact that subsoil use contracts have this specificity of their legal support, in comparison with the usual civil-law contracts, they at the same time retain the features inherent in the latter.

Analyzing the general legal specifics of energy contracts, we can conclude that the existing name of the subsoil use contract contains a legal collision, since in civil law there are provisions regulating relations in a simple contract and provisions governing the legal relationship arising from the license contract. Since, as shown by the study of the substance of the subsoil use agreement, this contract is licensed, in order to eliminate the identified conflict, it seems right to determine the unified name of this agreement, calling it a "license agreement on subsoil use".

A comparative analysis of various types of agreements in the field of subsoil use made it possible to conclude that the most acceptable and beneficial for the state in the conduct of oil operations is the conclusion of a production sharing agreement. In a comparative legal analysis with this type of contract as a concession agreement (Doronina 2014), the latter also consists in the process of the emergence of economic relations with respect to petroleum products of consumption. However, the license agreement has a wider range of activities and is not limited to "oil agreements" only.

We believe that in order to prevent the emergence of conflict situations between energy partner countries, the contract for the extraction, purchase or supply of energy products from Russia to a foreign country should be of an exact nature, that is, its main provisions relating to the order of supply and performance by the parties of their obligations under the contract should be prescribed in such a way that in the event of a violation, non-fulfillment or improper performance by a party of its obligations, this party would suffer material losses, however, when there is a conflict between the governments of states, there should not be a situation in which citizens of one of the countries become "hostages" to a political conflict.

An important place in the development of subsoil use is occupied by agreements on guarantees of foreign investments between states in the energy sector, among which the production sharing agreements (PSA) play an important role.

As the world practice shows, the most effective and mutually beneficial form of foreign investment in the development and extraction of mineral raw materials is the investment agreements that provide them to the parties—the state and investor —with maximum guarantees of proper performance of contractual obligations.

The main difference between the PSA contract and the state contract in the funding source: under the PSA, the work is financed directly by the investor, and only then the state reimburses the costs of the production.

The main source of legislative regulation in this area is Federal Law of 30.12.1995 No 225-FZ "On Production Sharing Agreements" (Federal Law 1995) (hereinafter referred to as "On the PSA"), which regulates the issues of concluding and terminating agreements, as well as Chapter 26.4 (articles 346.34–346.42) of the Tax Code of the Russian Federation—the legal basis for regulating the taxation of entities that have concluded an agreement, as well as the taxation of the extraction of minerals produced within the framework of the PSA.

The PSA develops as a civil-legal design—initially the parties to the transaction use civil-law methods for regulating relations arising in subsoil use.

Analyzing the legal nature and essence of the PSA agreement, it can be concluded that the production sharing agreement, by analogy with the subsoil use contract, is a civil law contract. One of the parties to the agreement is always the state, and its counterparty—the investor is a natural or legal person, undertaking to conduct exploration, extraction of minerals in order to obtain profits for the parties to the agreement (Marchukov 2016).

10   Conclusion

Summarizing the above, one can come to the conclusion that the peculiarities of the legal regulation of Russia's foreign trade contractual relations in the energy sector are primarily due to the specific nature of energy resources as a turnover object and a special commodity with specific properties.

Analysis of existing legal constructions of civil law nature applicable in the sphere of oil and gas turnover showed that they include: (1) the upstream contractual designs necessary to gain access to the search and development of oil and gas fields, as well as directly directed to exploration and development of oil and gas fields—oil service agreements; (2) midstream agreements related to transportation; (3) downstream agreements governing the processing and marketing. In general, in the energy sector—this is a transaction, the subject of which is the purchase and sale, transportation, transportation of energy resources, purchase and sale of power equipment; transactions, the subject of which is the design, construction of energy facilities, and others.

The contract for foreign trade in the sale of energy resources is a document that formalizes a foreign trade transaction, by virtue of which the seller is obliged to transfer energy resources to the buyer's property, and the buyer must accept this product and pay a certain amount of money for it.

It is established that when concluding foreign trade contracts in the sphere of international energy resources, the advantage is given to long-term contracts. Taking into account the made conclusion, and also with the purpose of elaboration of more detailed recommendations regarding the preparation and content of a long-term foreign trade contract (contract) for the sale and purchase (supply) of energy resources, it is possible to formulate recommendations on a list of essential and recommended, mandatory conditions. Their definition is given above, they should be included in the content of the foreign trade contract (contract) for the supply of energy resources. In addition to the essential for all types of international contracts of sale and purchase conditions, such as: the subject of the contract (name, quantity of goods) and price, the following must be attributed to the mandatory conditions of the foreign trade contract for the supply of energy resources: index-ation of prices; taxes, costs and expenses; conditions of payment; Scope of supply; product quality; place and time of delivery; acceptance of goods; guarantees; insurance; force majeure circumstances; ways to deal with disputes; responsibility of the parties; validity; applicable right; a condition that prevents unilateral termi-nation of contractual obligations, except in cases of prolonged force majeure circumstances.

In addition, the list of recommended as mandatory conditions for a long-term foreign trade contract for the supply of energy resources must contain specific conditions inherent to it due to special properties of the goods and technology for the implementation of contractual relations, such as: technical conditions; balancing of deliveries on requests; the condition "take-or-pay" ("take or pay"); mechanisms for ensuring flexibility, the systematization of shortages.

References

Abova, T. E., Belyaeva, Z. S., Gendzekhadze, E. N., et al. (2007). *Commentary on the civil code of the Russian federation, part one: In 3 tons (post-article)*. Moscow: Yurayt.

Agarkov, M. M., & Bratus, S. N. et al. (1944). *Civil law. Textbook. T.2*. The jurist. Publishing House of the NKJ USSR, Moscow.

Asoskov, A. V. (2012). *The basics of conflict law*. Moscow: Infopropic Media.

Asoskov, A. V. (2014). Reform of section VI private international law of the civil code of the Russian federation. *Economy and Law, 2*, 3–28.

Belov, V. A. (2013). *Civil law. T.II. A common part. Persons, blessings, facts: A textbook for bachelors.* Yurayt, Moscow.

Boguslavsky, M. M. (2016). *International private law: A textbook yur.* Moscow: Norma, SIC INFRA-M.

Braginsky, M. I., & Vitryansky, V. V. (1997). *Contract law: General provisions.* Moscow: Statute.

Braginsky, M. I., Klein, N. I., Levshina, T. L., & Litovkin, V. N. et al. (1997). *Civil law of Russia: Obligations. lecture course.* Part 2, BEK, Moscow.

Del Río, P. (2017). Why does the combination of the European Union Emissions Trading Scheme and a renewable energy target makes economic sense? *Renewable and Sustainable Energy Reviews, 74,* 824–834.

Dmitrieva, G. K. (2013). *Legal regulation of foreign economic activity in the context of Russia's accession to the World Trade Organization. Monograph,* Norma, Infra-M, Moscow.

Doronina, N. G. (2014). *Concession agreement in private international law. Certain types of obligations in private international law: Monograph,* INFRA-M: ISISP, Moscow.

Federal Law. (1995). *On production sharing agreements, (approved by the Decree of the Government of the Russian Federation of December 30, 1995 no. 225-FL).* (Russian).

Fedoseyeva, G. Y. (2005). *International private law.* Moscow: Textbook Eksmo.

Funk, Y. I. (2005). *International trade law: Contracts for the international sale of goods and international trade intermediation.* Minsk: Dikta.

Galkina, N. M. (2014). Peculiarities of the contract for the supply of oil and oil products. *Business in Law, 3,* 113–117.

Gas Forum. (2017). *Proposals on the organization of access to the gas transportation system of OAO Gazprom.* Retrieved March 22, 2017, from http://gasforum.ru/obzory-i-issledovaniya/1605/#5. (Russian).

Gazprom. (2017a). *About Gazprom.* Retrieved January 22, 2017, from http://www.gazprom.ru/about/marketing/europe/. (Russian).

Gazprom. (2017b). *Corporate magazine of PJSC Gazprom.* Retrieved January 22, 2017, from http://www.gazprom.ru/f/posts/44/119716/gazprom_12_2016_s.pdf. (Russian).

Gazprom in Questions and Answers. (2017). *Gazprom in foreign markets.* Retrieved January 22, 2017, from http://www.gazpromquestions.ru/foreign-markets. (Russian).

Gudkov, I. V. (2008). *Gas export and construction of trans boundary pipelines: Some aspects of legal and contractual regulation. Energy and law,* Lawyer, Moscow.

Inshakova, A. O. (2015). Foreign economic transactions in the updated civil legislation of the Russian Federation: qualification, form, applicable law. *Lawyer, 13,* 11–16.

Ioffe, O. S. (1975). *Liability,* legal literature, Moscow.

Ivanov, A. A. (2013). Oilfield services upstream agreements. The system of contractual relations between oil and gas companies in Russia at the stage of exploration and production of hydrocarbon raw materials. *Energy law, 2,* 45–48.

Kanashevsky, V. A. (2008). The notion of a foreign economic transaction in Russian law, doctrine and arbitration practice. *Journal of Russian Law, 8,* 74–76.

Kanashevsky, V. A. (2009). *International private law: A textbook.* Moscow: Intern. relations.

Kiyashko, V. A. (2007). *Failed transactions: An outline of legislation, theory and law enforcement practice.* St. Petersburg: Legal Center Press.

Klein, N. I. (1976). *Organization of contractual and economic relations.* Moscow: Jurid. Lit.

Kozlov, S. V. (2015). Production and economic structure and legal status of the subjects of the Russian natural gas market. *Actual Problems of Business Law: A Collection of Articles,* 118–132.

Krasavchikov, O. A. (1985). *Soviet civil law: Textbook: In 2 t.,* High school, Moscow.

Krylov, I. A., & Stolyarova, A. A. (2016). The main international agreements in the field of oil and gas transportation. *Economics, 45,* 207–217.

Kurlychev, D. V. (2008). Requirements for registration of foreign economic transactions. *Journal of Russian Law, 7,* 25–28.

Lakhno, P. G. Zekker, F. Y. (2011). *Energy law of Russia and Germany: Comparative legal research*, Publishing group Lawyer. Moscow.

Law R. U. (2017). *Sementsov, PO, contracts take-or-pay (take or pay)*. Retrieved March 22, 2017, from https://zakon.ru/discussion/2015/9/28/o_dogovorax_takeorpay_beri_ili_plati. (Russian).

Lunts L. A. (1973). *The course of private international law. A common part*, Legal literature. Moscow.

Makovsky, A. L., & Khlestova, I. O. (2012). *Problems of unification of private international law*. Moscow: ISISP Jurisprudence.

Marchukov, I. P. (2016). Production sharing agreement as a special type of civil contractual obligations in the energy sector. *Science and Education: Economy and Economy; Entrepreneurship; Law and Administration, 3*(70), 32–36.

Marysheva, N. I. (2000). *International private law: A textbook*. Moscow: Infra-M Contract.

Meyer, D. I. (1997). *Russian civil law (at 2 pm)*. Moscow: Statute.

Moroz, S. P. (2007). Oil contracts. *Power Law, 2*, 33–39.

Neshataeva, T. N. (2002). International commercial transactions: Legal regulation and judicial practice. *Arbitration and Civil Process, 6*, 24–34.

Novitsky, I. B., & Lunts, L. A. (1950). *General doctrine of obligations*. Moscow: Legal Literature.

Osmolovskaya, Y. A. (2013). The problem of implementing the principle of freedom of contract in determining the essential terms of the contract. *Business Security, 2*, 16–18.

Peace and Right. (2017). *International contracts for the supply of oil, oil products and natural gas*. Retrieved March 22, 2017, from http://www.miripravo.ru/contract/sale/oil/. (Russian).

Prokushev, E. F. (2012). *External Economic Activity: Textbook*, The publishing and trading corporation Dashkov and Co. Moscow.

Puginsky, B. I. (2016). *Selected works: A collection for the 75th anniversary*. Moscow: Yurayt.

Rodionova, I. A., Chernyaev, M. V., & Korenevskaya, A. V. (2017). Energy safety and innovative development of the BRICS States. *International Journal of Energy Economics and Policy, 7* (3), 216–224.

Romanova, V. V. (2016). Peculiarities of legal regulation of foreign economic transactions in the energy sector. *International Public and Private Law, 1*, 16–21.

Rosenberg, M. G. (2004). *The contract of international sale and purchase: Commentary to the legislation and practice of dispute resolution*. Moscow: Statute.

Rosenberg, M. G. (2006). *International sale of goods: A commentary on legal regulation and dispute resolution practices*. Moscow: Statute.

Rosneft. (2017). *Rosneft and PVOIL have concluded an unprecedented long-term contract for the supply of oil*. Retrieved January 22, 2017, from https://www.rosneft.ru/press/releases/item/182553/. (Russian).

Sadikov, O. N. (1986). *Conflict rules in private international law*. Moscow: Legal Literature.

Seinaroyev, B. M. (2000). The energy supply contract. *Bulletin of the Supreme Arbitration Court of the Russian Federation, 6*, 128–141.

Semenikhin, V. V. (2015). *External economic activity*. Moscow: GrossMedia ROSBUKH.

Sergeev, A. P., & Tolstoy, Y. K. (2000). *Civil law. Volume 1. The textbook*, PBOYUL L.V. Rozhnikov. Moscow.

Shestakova, M. P. (2014). *The contract of international sale and purchase (sources and principles of legal regulation)*. Moscow: INFRA-M.

Shevchenko, L. I. (2013a). The problems of determining the essential conditions for the conclusion of contracts in the sphere of the fuel and energy complex. *Law and Economics, 6*, 29–35.

Shevchenko, L. I. (2013b). Development of legislation regulating oil transportation relations in the system of main oil pipelines and its impact on theoretical concepts of their legal nature. *The Legal World, 5*(197), 14–19.

Sinaisky, V. A. (1917). *Russian civil law*. Kiev: Progress.

Sukhanov, E. A. (2000). *Civil law: B.2 vol. Volume II. Semitic 1: Textbook*, BEC, Moscow.

Sukhanov, E. A. (2008). *Civil law. In 4 volumes. Volume 3: Obligations: Textbook*, Wolters Clover, Moscow.

Vilkova, N. G. (2002). *Contract law in international circulation*. Moscow: Statute.

Vitryansky, V. V. (1999). *The contract of sale and its separate varieties*. Moscow: Statute.
Vlasov, A. A., & Kovalenko, V. N. (2015). Regulation of foreign economic activity by the rules of private international law. *International Public and Private Law*, *1*, 20–24.
Yurchenko, A. K. (1961). *Soviet Civil Law. Course of lectures: Certain types of obligations: Textbook*, Leningrad State University, Leningrad.
Zeng, S., Liu, Y., Liu, C., & Nan, X. (2017). A review of renewable energy investment in the BRICS countries: History, models, problems and solutions. *Renewable and Sustainable Energy Reviews, 74,* 860–872.
Zvekov, V. P. (2004). *International private law: A textbook*. Moscow: Lawyer.

Applicable Law as an Essential Condition for a Foreign Trade Contract for the Supply of Energy Resources

Agnessa O. Inshakova and Igor P. Marchukov

1 Annotation

The chapter is devoted to the study of modern legislative and doctrinal approaches to determining the applicable law, as a condition for a foreign trade contract for the supply of energy resources.

It is noted that issues related to the definition of the applicable law to foreign trade contracts for the supply of energy resources are becoming particularly relevant.

This is due to the transnational nature of the movement of goods and involves participation in commercial legal relations of a foreign element associated with the presence in their regulation of another legal system with a high cost of goods and the risk of causing major damage to the party as a result of improper performance of obligations.

The problems arising in connection with the lack of uniformity of approaches in theory and jurisprudence to the scope and interpretation of the category "applicable law" are disclosed. The conclusion is made about the expediency of broad interpretation of this condition in relation to the content of the foreign trade contract for the supply of energy resources. It is also concluded that, established on the basis of the choice of the contract of supply by the parties, their own binding statute should take into account and rely on the whole set of norms of national law, norms and universally recognized principles of international law, as well as subsidiary

A. O. Inshakova (✉)
Department of Civil and International Private Law, Institute of Law,
Volgograd State University, Volgograd, Russia
e-mail: gimchp@volsu.ru; ainshakova@list.ru

I. P. Marchukov
Department of Civil and International Private Law, Volgograd State University,
Volgograd, Russia
e-mail: gimchp@volsu.ru

© Springer International Publishing AG, part of Springer Nature 2019
O. V. Inshakov et al. (eds.), *Energy Sector: A Systemic Analysis of Economy,
Foreign Trade and Legal Regulations*, Lecture Notes in Networks and Systems 44,
https://doi.org/10.1007/978-3-319-90966-0_4

non-state legal regulation in the form of customs and customs, traditionally applied in the field of international trade.

In this regard recommendations are proposed to improve the wording of paragraph 1 of Article 1186 of the Civil Code of the Russian Federation.

In the course of the study, differences in theory and law enforcement practice were revealed, related to the lack of legislative specification of the form of registration the terms of the contract and circumstances that testify to the will of the parties regarding the choice of the applicable law in the norm of clause 2 of Article 1210 of the Civil Code of the Russian Federation. It was established that the definition of the applicable law in the absence of legislative specification any recommendations developed by law enforcement practice, presents a certain complexity. In this regard, a recommendation is recommended in concluding foreign trade transactions in the energy sector to express an agreement on the choice of the law to be applied by direct indication of this in the contract.

2   Materials

The normative basis of the study was the provisions of positive law contained in the national acts, namely, the Constitution of the Russian Federation, Sect. 6 of the Civil Code of the Russian Federation, the Arbitration Procedure Code of the Russian Federation, the Federal Law of the Russian Federation "On International Treaties of the Russian Federation", as well as in international treaties, in particular, in the Rome I Rules, the Rome Convention of 1980 and the 1994 Inter-American Convention. The legal basis for the research was also the codes of commercial customs (Lexmercatoria), first of all, the rules for the interpretation of the international terms of Incoterms—2010 and the Principles of International Commercial Contracts—Unidroit.

The concept of the applicable law in the foreign trade contract for the purchase and sale of energy resources and approaches to determining its volume have been studied using the example of works by Garagulya (2010), Strigunova (2016). Theoretical approaches to determining which contract terms and circumstances indicate the choice of the applicable law by the parties were studied using the example of Vlasov and Kovalenko (2015).

The doctrinal analysis of judicial and arbitration practice concerning use by the parties of the contract of terminology and formulations of the legislation of the Russian Federation as a held choice of the applicable law was made with the recognition of the opinions expressed in the scientific literature Filonova and Kruchinin (2014).

The empirical basis of the research is based on an analysis of the position of the RF courts on the definition of the scope of the notion of applicable law, in particular, the Information Letter of the Presidium of the Supreme Arbitration Court of the Russian Federation of 09.07.2013 No. 158 "Review of judicial practice on certain issues related to the consideration by the Arbitration Courts of cases involving foreign persons", Decision of the Arbitration Court of the Novosibirsk Region on November 18, A45-14953/2016, Attributive of the Arbitration Court of the Chelyabinsk region of December 23, 2016 in case No. A76-31110/2016; Attributive of the Arbitration Court of the Volgograd Region of 28.09.2010 in case No. A12-2488/2010, Prescript of the Fifteenth Arbitration Appeal Court of August 26, 2015, No. 15AP -13087/2015 in case No. A32-12037/2015.

3 Methods

Methods of synthesis, classification, generalization, formal-logical, dialectical, system-structural, cognition, empirical description were used in the research process as general scientific methods. The private scientific methods used in the research include: formal legal, the principle of evaluating legal processes, comparative legal, etc. The procedure and conditions for concluding a foreign trade contract in the field of energy supplies are examined using the structural-functional analysis method. The conflicts arising from the implementation of the parties of the contract were investigated using a process-dynamic method.

4 Introduction

In the context of the study of the essential terms of a foreign trade contract for the supply of energy resources, issues related to the definition of the applicable law acquire special urgency, since energy resources are primarily a transnational product, which implies the participation of foreign elements in trade relations and the possibility of another legal system being in their regulation.

In Russian and international legislation there is no legal definition of the term "applicable law", which causes a large number of disagreements in theory and practice regarding its substance and legal nature. In the scientific literature, there are several approaches to determining the scope of this concept. Thus, most authors believe that the scope of the term "applicable law" should be limited to the framework of national law applicable to the legal relations of the parties to a foreign trade contract. The main argument in favor of this point of view is the fact that international law, as such, is not an independent body of legal norms regulating foreign trade contractual relations without involving national law. Thus, in accordance with paragraph 4 of article 15 of the Constitution of the Russian Federation, generally recognized principles and norms of international law and international treaties are included in the legal system of the Russian Federation (The Constitution of the Russian Federation 1993; Federal Law 1995). This provision is also reflected in Article 7 of the Civil Code of the Russian Federation.

However, we believe that we should agree with the researchers, who believe that the term "applicable law" (lex contractus) is used in the modern Russian legislation in the narrow sense (Garagulya 2010). Systemic interpretation of the provisions of the Civil Code allows us to conclude that the legislator includes in the category of applicable law, the right of a state, subject to application in the relations conditioned by the foreign trade agreement. The supply of energy resources, primarily oil and petroleum products, is often carried out through the use of merchant ships of the tanker type, in this connection, attention should be drawn to Article 414 of the Merchant Shipping Code of the Russian Federation, according to which, in the category of applicable law, along with the national legislation of the Russian Federation, international treaties and customs of merchant shipping, the so-called lex mercatoria, are also included (Backer 2017). Let's try to understand.

5   Elucidation of the Scope of the Concept of Law Applicable to Foreign Trade Contracts for the Sale and Purchase of Energy Resources

The difference in the approaches of the legislator to determining the scope of the term "applicable law" indicates a lack of uniformity in the solution of this issue. In this case, we should agree with the point of view of D.P. Strigunova, who believes that the term "applicable law" should be interpreted broadly, because the parties to the treaty have the option of choosing not only the national law to be applied, but also various legal systems (Strigunova 2016). The Constitution of the Russian Federation in article 7 indicates that the Russian legal system, along with international treaties, includes generally recognized principles of international law. In this connection, it is natural to ask whether the universally recognized principles of international law, customs and practices of international trade are included in the scope of the term "applicable law".

Before answering this question, one should turn to the legal nature of the commercial custom and the general principles of law. These legal categories exist on the basis of the lex mercatoria doctrine and are widely used in European countries. The essence of this doctrine is that international trade relations should be regulated primarily by commercial contracts and international trade practices.

In the sphere of international trade in energy resources, as already mentioned, two sets of trade customs can be distinguished, these are the rules for the interpretation of the international terms Incoterms 2010 and the Principles of International Commercial Contracts Unidroit, which are the brightest example of Lex mercatoria (Chung and Lee 2013). It should be noted that despite the convenience in the application and the plasticity of the provisions, trade customs are not independent rules of law, because they are not sanctioned by the state. Thus, commercial customs can be included by the parties to the contract in the scope of the applicable law, but not as an independent source of legal norms, but in the form of a non-state legal regulator that is applied in a subsidiary manner with the norms of international and national law.

6　Case Study

The position of the courts of the Russian Federation on the issue of determining the scope of the concept of applicable law is set forth in the Information Letter of the Presidium of the Supreme Arbitration Court of the Russian Federation of 9.07.2013, according to which, if the general principles of law, commercial customs and customs, other sources of Lex mercatoria are indicated in the contract as an applicable law, the state courts of the Russian Federation do not recognize such a choice as due to the fact that the parties to the treaty can choose exclusively national law (Codification of the Russian Federation 2017). When this case appears, in practice, the applicable law is established by the courts of the Russian Federation by using conflicts of law rules.

Thus, summarizing the analysis of research approaches (Garagulya 2010; Strigunova 2016), positions of legislation and courts of the Russian Federation, we believe we should adopt an expanded interpretation of the category "applicable law", which allows to include in its scope the totality of universally recognized principles and norms of international law, the norms of national law, as well as subsidiary non-state legal regulation of relations of counterparts in foreign trade contracts in the form of established trade customs and customs.

7　Specifics of Establishing the Category of Applicable Law in the Foreign Trade Contract for the Supply of Energy Resources

Let us analyze the features of establishing the category of applicable law in the foreign trade agreement for the supply of energy resources in accordance with the legislation of the Russian Federation. The issues of determining the applicable law for a foreign trade contract for the supply of energy resources are resolved through the norms of Sect. 6 of the Civil Code of the Russian Federation. As a general rule, Russian, foreign physical and legal persons, as well as stateless persons, can enter into contractual relations of this type. When determining the law applicable to a foreign trade contract, the principle of autonomy of the will of the parties operates (Yerpyleva 2015; Kuznetsov 2015). It consists in the fact that the parties to the agreement are free to choose not only its content, but also the type of legal norms. This principle received its legislative implementation in Article 1210 of the Civil Code of the Russian Federation, according to which, the parties to the contract may, at the time of entering into a contract or subsequently choose by agreement between themselves, which is subject to application to their rights and obligations under this treaty. It is interesting that the agreement on the choice of the applicable law is retroactive, and it can also act both in relation to the entire treaty and in respect of certain of its provisions. These opportunities provided to the parties to the contract are of great importance in law enforcement practice. The foreign trade contract for the supply of energy resources requires detailed study of all its items, as it implies a high cost of the goods, and the risk of causing major damage to the party as a result of improper performance of obligations. In this regard, the provision of opportunities for the parties to determine the applicable law to certain clauses of the treaty is of particular relevance.

The requirement of the legislator to the form of the agreement is dispositive. So, according to Part 2 of Article 1210 of the Civil Code of the Russian Federation, the parties have three options for expressing their will regarding the choice of the applicable law: by direct expression in the contract (addition), by indirect expression, however, in such a case, the terms of the contract or the circumstances of the case in their totality should testify to the will of the parties (Garagulya 2010).

A similar wording on the form of the agreement of the parties on the choice of the law to be applied is fixed in Art. 3 (1) of the Rome I Convention, the Rome Convention of 1980, the Inter-American Convention of 1994. It was also widely accepted in international treaties and national legislation of various countries.

The provision of Part 2 of Article 1210 of the Civil Code of the Russian Federation, which appeared in the part of the third Civil Code of the Russian Federation, enacted since March 1, 2002, was a novelty of the Russian legislation and was developed on the basis of the provisions of the 1980 Rome Convention. Relatively fuzzy legislative formulation that the agreement of the parties on the choice of law should definitely follow from the terms of the contract or the totality of the circumstances of the case is more typical for countries of common law than for continental Europe. At the same time, researchers note that foreign and international practice of using exactly such "flexible" formulations testifies to their effectiveness (Garagulya 2010).

The choice by the parties to the contract of the applicable law means the establishment of its own binding statute (Veselkova 2013). Since the conclusion of the agreement, the counterparties undertake to approach the interpretation of the foreign trade agreement, in part not regulated by its terms, from the position of the legal system chosen by them. It should be noted that the principle of autonomy of the will of the parties, established by article 1210 of the Civil Code of the Russian Federation, has a limitation of scope. This situation develops when the mandatory norms of Russian legislation contain an indication of their particular importance for securing the rights and interests of the participants in civil circulation, then they regulate such relations regardless of the terms of the agreement of the parties and conflict rules. Such provisions of the Russian legislation are called norms of direct application. As such a norm of direct application it is possible to cite the wording of clause 3 of article 162 of the Civil Code of the Russian Federation, effective before 1.09.2013, according to which non-compliance with a simple written form of an external economic contract entails the invalidity of the transaction.

The lack of a proper specification of the norms of part 2 of article 1210 of the Civil Code of the Russian Federation on the way to formalize the will of the parties with regard to the choice of the applicable law raises many disagreements in law enforcement practice. Thus, the legislator does not specify which contract terms and circumstances indicate the choice of the applicable law by the parties. According to

A.A. Vlasov, V.N. Kovalenko, the above issues should be resolved by the court within the framework of each individual case, proceeding from the circumstances of execution (default) of obligations and conditions of the contract (Vlasov and Kovalenko 2015).

8 Arbitration Practice of the Russian Federation in Questions of Choice by the Parties of the Law Applicable to Contracts for the Supply of Energy Resources

In the arbitration practice of the courts of the Russian Federation, the question of the choice by the parties of the applicable law does not arise only in the case of the direct expression of the will of the parties by including the relevant clause in the terms of the contract.

9 Case Study

So, in the Decision of the Arbitration Court of the Novosibirsk Region on 18.11.2016, the Court settles the dispute under the supply contract with the application of the substantive law of the Russian Federation on the basis of the contract item, according to which "the law governing the contract is the legislation of the Russian Federation". At the same time, the Arbitration Court of the Novosibirsk Region did not agree with the opinion of the defendant—the Russian organization on the existence of grounds for termination of the proceedings in the case and the transfer of the statement of claim to the competent court of the Republic of Turkey, since the place of execution of the contract, based on the basis of supply, is Mersin, Turkey (The Electronic Bank of the Arbitration Courts 2017f).

In order to form a uniformity of judicial practice and prevent violation of the right to appeal to the court, provided for in article 4 of the Arbitration Procedure Code of the Russian Federation (Arbitration Procedural Code of the Russian Federation 2002), the Supreme Arbitration Court of the Russian Federation prepared an Information Letter with explanations regarding the parties' choice of foreign trade relations of the applicable law.

Thus, in the opinion of the Supreme Arbitration Court of the Russian Federation, the choice by contractors of a particular court of a state as an authority authorized to deal with disputes arising in the context of their contract does not in itself imply a choice as applicable, the law of the state where the court is located (Codification of the Russian Federation 2017). The determination by the parties to the contract of the place of dispute settlement by a court in the territory of the Russian Federation means the empowerment of the said court with powers to determine the applicable law, taking into account conflict rules, but no more (Vlasov and Kovalenko 2015). The Russian-speaking nature of the foreign trade contract for the supply of energy resources, its conclusion on the territory of Russia, also, in the opinion of the Supreme Arbitration Court of the Russian Federation, can not testify to the choice by the parties to the contract as the applicable right of substantive law of the Russian Federation. When determining the applicable law to the foreign trade contracts for the supply of energy resources, attention should be paid to the following peculiarities associated with the specific nature of the delivery objects—energy resources and energy carriers. Thus, the choice of the applicable law by the parties to the contract can't be proved by the conditions of the country of origin, extraction of energy resources, the location of the shipment/unloading, the greater length of the main pipeline, the country of development of the applied standard (GOST) of the quality of the supplied energy resource, etc.

The system analysis of the review of the judicial practice of arbitration courts on some issues related to the consideration of cases involving foreign persons allowed to identify a number of circumstances, in the aggregate of which, the court recognizes the parties' choice of the contract of the applicable law held. In particular, this circumstance may be the justification by the parties, during the judicial process of their demands and objections, with reference to the same applicable law (Codification of the Russian Federation 2017).

This provision also corresponds to paragraph 2 of the Information Letter of the Russian Supreme Arbitration Court of 9.07.2013 No 158, according to which the Arbitration Court recognizes as legitimate the conclusion of the so-called prophecy agreement, on which the parties agree that disputes must be considered in the court of the country by one of them, which in the future will appear in court as a plaintiff or a defendant (Codification of the Russian Federation 2017).

10 Case Study

An example of recognition by the Arbitration Court of the Russian Federation of a prophecy agreement may be the Definition of the Arbitration Court of the Chelyabinsk Region of 23.12.2016, in which the court points to a literal interpretation of the terms of the two contracts, which indicates that the parties have provided a prophecy agreement and the competent court is arbitration at the location of the claimant, the substantive law liable to be applied is the law of the Russian Federation (Electronic Bank of Arbitration Courts decisions 2017b). A similar position can be traced in the definition of the Arbitration Court of the Volgograd Region in the case of recovery of debts under a foreign trade contract for the supply of petroleum products between PJSC Lukoil-Inter-Card and UAB LINAVOS SERVISAS (Lithuania) (Electronic Bank of Arbitration Courts Decisions 2017a). It should be noted that the use by the parties of a foreign trade contract for the supply of energy resources of a prophecy agreement in practice may have a number of negative legal consequences. Thus, the parties are unable to resolve the dispute by mediation procedure, since before the time of referral to the court, the prophecy agreement is invalid, and there is no possibility to determine the right of which party to the contract is to be applied. Such a risk of uncertainty of the applicable law can be traced in a number of judicial acts.

The Arbitration Courts of the Russian Federation recognize the use by the parties of the contract of the terminology and formulations of the legislation of the Russian Federation as a held choice of the applicable law. This approach of judicial practice is supported by the authors in the scientific literature. So, in A.A. Filonova, V.N. Kruchinin, the reference of the parties to a foreign trade contract for the current regulatory and legal acts of the Russian Federation means that Russian law is chosen as applicable to the specified contract (Filonova and Kruchinin 2014).

The agreement of the parties to a foreign trade agreement on the supply of energy resources as an applicable law of the Russian Federation indicates the choice of the entire Russian legal system, which, along with domestic law, includes international legal acts relating to conflict-of-laws regulation. In accordance with Article 5 of the Federal Law "On International Treaties of the Russian Federation", international treaties are an integral part of the legal system of the Russian Federation and have priority over national legislation (Electronic Bank of Arbitration Courts Decisions 2017c, 2017d). This norm indicates the importance of applying international legal norms.

In the process of developing and concluding a foreign trade contract for the supply of energy resources, the parties do not always seek to exercise the right provided for in article 1210 of the Civil Code of the Russian Federation. According to article 1211 of the Civil Code of the Russian Federation, in the absence of an agreement of counterparties on the applicable law to the foreign trade contract, the law of the country where, at the time of the conclusion of the contract, the place of residence or principal place of business of the party is applied, which carries out the fulfillment of the obligation, which is of the most decisive importance for the content of the contract. For each separate type of contractual obligations, a definition of the party is established, the existence and activities of which are most closely related to the contract. So, for the contract of international supply of energy resources, the organization of the seller is of decisive importance, which is reflected in part 2 of article 1210 of the Civil Code of the Russian Federation. However, this rule does not have a super-operative nature, and does not apply if from the law, the terms of the contract, or a combination of established circumstances of the case, it follows that the contract is much more closely related to the law of another country than that indicated by the norm of part 2. In such a situation, the law of the buyer's country may be applicable to the contract.

In arbitration practice, courts define as the applicable law the legal system of the buyer's country in the presence of circumstances directly indicating the existence of a close relationship between the contract and the buyer (Electronic Bank of Arbitration Courts Decisions 2017e). In accordance with the selected parties or the applicable law determined in accordance with the norms of the Civil Code of the Russian Federation, the issue of the limitation period for foreign economic contracts of supply and the procedure for its application are being decided. Issues, related to the statute of limitations and the form of the contract when concluding and executing a foreign trade supply contract are closely related to the personal law of the legal entity. Under the legislation of the Russian Federation, the personal law of a legal entity is based on the conflict principle of incorporation, the essence of which is the law of the country in which the legal entity is registered, as Article 1202 of the Civil Code of the Russian Federation directly specifies (The Civil Code of the Russian Federation (Part Three) 2001; Abova and Boguslavsky 2004). A similar principle applies to organizations entering into trade relations with applicable Russian law, but not being legal entities under the legislation of a foreign state. The personal law of a legal entity is governed by all matters related to its legal capacity, creation, reorganization and liquidation processes, corporate relations between a legal entity and its founders.

11 Conclusion

Thus, issues related to the definition of the applicable law for foreign trade contracts for the supply of energy resources, which is due to the transnational nature of the movement of goods and involves participation in commercial legal relations of a foreign element associated with the presence in their regulation of another legal system, the high cost of goods and the risk of causing major damage to the party as a result of improper performance of obligations.

In accordance with the legal nature and substance defined in the course of the study, the category "applicable law", established on the basis of its choice by the parties to a foreign trade contract, its own binding statute of the parties must take into account and rely on the totality of the norms of national law, norms and universally recognized principles of international law, as well as subsidiary non-state legal regulation in the form of customs and customs, traditionally applied in the field of international trade (Ariffin and Yaakub 2017).

In this connection, we propose to state item 1 of article 1186 of the Civil Code of the Russian Federation in the following wording:

"1. The law applicable to civil-law relations involving foreign citizens or foreign legal entities or to civil-law relations complicated by another foreign element, including in cases where the object of civil rights is abroad, is determined on the basis of universally recognized principles and norms of international law, international treaties of the Russian Federation, this Code, other laws (point 2, article 3), as well as customs and customs recognized in the Russian Federation.".

Disagreements in theory and law enforcement practice are revealed, connected with the absence of legislative concretization of the way of registration, the terms of the contract and the circumstances evidencing the will of the parties regarding the choice of the applicable law in the norm of part 2 of article 1210 of the Civil Code of the Russian Federation.

The legislator, in clause 2 of Article 1210 of the Civil Code of the Russian Federation, provides the parties to the contract with an alternative in choosing the form of the agreement of the parties on the choice of the applicable law: or the presence of the expressed will of the parties (express), or the terms of the contract, or the circumstances of the case (tacitly expressed).

When the agreement on the applicable law is contained directly in the text of the contract (foreign trade contract) or in a separate document, it is not difficult to establish the existence of an agreement on the applicable law.

The "silently expressed will of the parties" confronts judges with the difficult task of determining the actual will of the parties to the treaty. In interpreting in this case, the subjective factor is of fundamental importance, which leads to a wide range of possibilities for the exercise of judicial discretion, unpredictability of the judge's withdrawal in relation to the applicable law and the uncertainty of the result for the parties.

Such a way of determining the applicable law in the absence of legislative concretization of the method of execution, the terms of the contract and the circumstances evidencing the will of the parties with regard to the choice of the applicable law in the norm of part 2 of Article 1210 of the Civil Code of the Russian Federation any recommendations developed by law enforcement practice, presents a certain complexity.

In order to avoid the abovementioned negative situations for the parties to the contract, we deem it expedient, as a recommendation in concluding foreign trade transactions in the energy sector, to agree on the choice of the applicable right to express by direct indication of this in the contract.

The condition on the applicable law is recommended to include in the content of the foreign trade contract the supply of energy resources along with other essential conditions for it.

References

Abova, T. E., Boguslavsky, M. M., et al. (2004). *Commentary on the civil code of the russian federation, part three.* Moscow: Yurayt.

Arbitration Procedural Code of the Russian Federation. (2002). *(Approved by the Federation Council on July 10, 2002 no. 95-FL).* (Russian).

Ariffin, A., & Yaakub, N. I. (2017). Unidroit principles of international commercial contract as the rules of law governing cross border contracts. *Advanced Science Letters, 23*(1), 478–481.

Backer, L. C. (2017). A lex mercatoria for corporate social responsibility codes without the state? A critique of legalization within the state under the premises of globalization. *Indiana Journal of Global Legal Studies, 24*(1), 115–146.

Chung, J. H. & Lee, B. S. (2013). A genealogical approach to the Incoterms rules and revised Incoterms 2010. *Journal of Korea Trade, 17*(2), 1–19

Codification of the Russian Federation. (2017). *Information letter of the Presidium of the Supreme Arbitration Court of the Russian Federation of 09.07.2013 № 158 Review of judicial practice on certain issues related to the consideration of cases by arbitration courts involving foreign persons.* Retrieved January 23, 2017, from http://rulaws.ru/vs_rf/Informatsionnoe-pismo-Prezidiuma-VAS-RF-ot-09.07.2013-N-158/. (Russian).

Electronic Bank of Arbitration Courts Decisions. (2017a). *Definition of the Arbitration Court of the Volgograd Region of 28.09.2010 in case No. A12-2488/2010.* Retrieved January 23, 2017, from http://kad.arbitr.ru/PdfDocument/6d9df9d6-430b-4a25-bd8f-7d7ab0c3aeed/A12-2488-2010_20100928_Opredelenie.pdf. (Russian).

Electronic Bank of Arbitration Courts decisions. (2017b). *Definition of the Arbitration Court of the Chelyabinsk region of December 23, 2016 in case No. A76-31110/2016.* Retrieved January 23, 2017, from http://kad.arbitr.ru/PdfDocument/b9087a6c-fc67-4db6-8180-adb88762a700/A76-31110-2016_20161223_Opredelenie.pdf. (Russian).

Electronic Bank of Arbitration Courts Decisions. (2017c). *Decision of the Nineteenth Arbitration Appeal Court of 31.03.2014 in case No. A08-807/2013.* Retrieved January 23, 2017, from http://kad.arbitr.ru/PdfDocument/7ce1b0c7-6f8a-4532-b0e0-f1444d0c1519/A08-8072013_20140331_Reshenija%20i%20postanovlenija.pdf. (Russian).

Electronic Bank of Arbitration Courts Decisions. (2017d). *Decision of the Fifteenth Arbitration Appeal Court of August 26, 2015, No. 15AP -13087/2015 in case No. A32-12037/2015.* Retrieved January 23, 2017, from http://kad.arbitr.ru/PdfDocument/fdc31e86-f85e-47e3-9a55-b0a11e7116a6/A32-12037-2015_20150826_Postanovlenie%20apelljacionnoj%20instancii.pdf . (Russian)

Electronic Bank of Arbitration Courts Decisions. (2017e). *Decision of the Arbitration Court of the Kaliningrad Region on 09.11.2015 in case No. A21-3809/2015.* Retrieved January 23, 2017, from http://kad.arbitr.ru/PdfDocument/bef7cac8-bad9-48ae-9261-143c39e05d0d/%D0%9021-3809-2015__20151109.pdf. (Russian).

Federal Law. (1995). *On International Treaties of the Russian Federation, (approved by the Decree of the Government of the Russian Federation of July 15, 1995 no. 101-FL).* (Russian).

Filonova, A. A., & Kruchinin, V. N. (2014). Specifics of determining the applicable law in the resolution of arbitration by courts of disputes with foreign participation. *Research Publications, 6*(10), 26–33.

Garagulya, M. I. (2010). The concept of the applicable law in foreign economic contractual relations. *The Space of the Economy,* (2–3), 217–221.

Kuznetsov, M. N. (2015). *International Private Law. General part: Lectures delivered at the Peoples' Friendship University of Russia in 2004–2014: textbook,* Russian University of Friendship of Peoples, Moscow.

Strigunova, D. P. (2016). Category applicable law in international treaties. *International Public and Private Law, 4,* 29–32.

The Civil Code of the Russian Federation (Part Three). (2001). *(Approved by the Federation Council on November 14, 2001 no.146-FL).* (Russian).

The Constitution of the Russian Federation. (1993). *(was adopted by popular vote on December 12, 1993).* (Russian).

The Electronic Bank of the Arbitration Courts. (2017f). *The decision of the Arbitration Court of the Novosibirsk Region on November 18, 2016 in case No. A45-14953/2016.* Retrieved January 23, 2017, from https://kad.arbitr.ru/PdfDocument/0d5e3c8b-2732-46d8-b231-9717acf4ad8a/A45-14953-2016_20161118_Reshenie.pdf. (Russian).

Veselkova, E. E. (2013). Conflict and substantive regulation of the contract of international sale. *Counsel, 8,* 9–12.

Vlasov, A. A., & Kovalenko, V. N. (2015). Regulation of foreign economic activity by the rules of private international law. *International Public and Private Law, 1,* 20–24.

Yerpyleva N. Y. (2015). *International Private Law: A Textbook for Universities,* Ed. house of the Higher School of Economics, Moscow.

Part II
Factors that Determine the Development of Foreign Trade Activities in the Energy Sphere and of its Economic-Legal Regulation

Innovation as an Integral Condition for the Development of Modern Foreign Trade Turnover of Energy Resources and its Legal Regulation

Agnessa O. Inshakova and Alexander I. Goncharov

1 Annotation

Taking into account the realities of modern world economic development—the transition of leading countries to the construction of an economy, the foundation of which is the creation, dissemination and use of knowledge, special attention is paid to studying one of the most important factors affecting the development of foreign trade turnover of energy resources—the level of innovation in the energy sector.

In all developed countries, various measures are being taken to support innovation. Innovative activity in the field of energy gives hope for resolving one of the priority problems of the 21st century—the depletion of energy resources. In this chapter we analyze the documents of strategic planning in the field of innovative development in the energy sector: Strategy of innovative development of the Russian Federation for the period until 2020 the Energy Strategy of Russia for the period up to 2030 the draft of the Energy Strategy of Russia for the period up to 2035, Plan of measures ("road map") "Introduction of innovative technologies and modern materials in the fuel and energy sector" for the period until 2018, Forecast of scientific and technological development of the fuel and energy sector of Russia for the period up to 2035. Also the programs of innovative development of the largest energy companies are being studied—PJSC Gazprom and PJSC NK Rosneft.

The authors substantiates the conclusion that the innovation activity consists in the introduction of the newest highly efficient technologies and equipment, the use of advanced world experience, the improvement of the ecological character of

A. O. Inshakova (✉) · A. I. Goncharov
Department of Civil and International Private Law, Institute of Law,
Volgograd State University, Volgograd, Russia
e-mail: gimchp@volsu.ru; ainshakova@list.ru

A. I. Goncharov
e-mail: gimchp@volsu.ru; goncharova.sofia@gmail.com

© Springer International Publishing AG, part of Springer Nature 2019   67
O. V. Inshakov et al. (eds.), *Energy Sector: A Systemic Analysis of Economy,*
Foreign Trade and Legal Regulations, Lecture Notes in Networks and Systems 44,
https://doi.org/10.1007/978-3-319-90966-0_5

production and the level of processing of raw materials the transition to modern types of raw materials and fuels and the development of energy based on the use of alternative and renewable sources energy.

2 Materials

The normative basis for the study was mainly strategic acts of strategic nature defining the state policy, the main tasks and principles of implementation, the priority directions for the development of the energy sector and the activities needed in these areas.

Among them above all, the Energy Strategy of Russia for the period up to 2030, approved by the Decree of the Government of the Russian Federation No. 1715-r of November 13, 2009, the Concept of Long-Term Social and Economic Development of the Russian Federation for the period to 2020, approved by the Decree of the Government of the Russian Federation of November 17, 2008 № 1662-r, The Strategy of Innovative Development of the Russian Federation for the Period to 2020, approved by the Decree of the Government of the Russian Federation No. 2227-r 08.12.2011, Plan of measures ("Road Map") "Introduction of innovative technologies and modern materials in the fuel and energy sector" for the period until 2018, approved by the Decree of the Government of the Russian Federation No. 1217-r dated July 3, 2014, The federal target program "Research and development in priority areas of development of the scientific and technological complex of Russia for 2014–2020", approved by the Decree of the Government of the Russian Federation No. 426 of May 21, 2013, Forecast of scientific and technological development of the fuel and energy sector of Russia for the period up to 2035, The draft Energy Strategy of Russia for the period up to 2035, taking into account the results of the annual monitoring of the implementation of the Energy Strategy of Russia for the period until 2030.

The concept of innovative activity and its qualitative characteristics as a kind of entrepreneurship in the economic and legal aspect are defined in the work based on the works Gribanov (2016), Inshakova and Ryzhenkov (2015), Frolov (2017).

The study of the necessary conditions for the development of legal support for innovation in the field of energy is studied on the basis of the works of Romanova V.V.

The empirical basis of the research was made by local regulations and official websites of the largest oil and gas companies that are geared to global development as a reliable energy supplier and take into account the need to solve various tasks that require the search, acquisition and application of new knowledge, the continuous increase in activity and efficiency of innovation. Among them are such documents as the passport of the Innovative Development Program of PJSC Gazprom until 2025, the Program of Innovative Development of PJSC NK Rosneft the

Passport of the Innovative Development and Technological Modernization Program of Rosatom State Corporation for the period up to 2030 (in the civilian part), Portal of innovative cooperation of PJSC "LUKOIL".

3 Methods

In the process of carrying out the research, both general scientific methods (formal-logical, dialectical, system-structural, analysis, induction, deduction, synthesis and generalization) were used, as well as private-science methods (formal-legal, legal process, comparative-legal).

4 Introduction

The Russian Federation as one of the world's major exporters of energy resources is extremely interested in ensuring further enhancement of production efficiency and expansion of exports of all major energy resources, products of their processing, as well as high technologies in which Russian energy and industrial companies have competitive advantages. Among the main problems that negatively affect the development of foreign trade turnover of energy resources and its effectiveness are:

- low level of presence of Russian energy companies in foreign markets;
- lower prices and reduced demand for energy due to the global financial and economic crisis;
- maintaining the dependence of Russian exports on transit countries;
- weak diversification of the markets for the sale of national energy resources and the commodity structure of exports;
- politicization of Russia's energy relations with foreign countries.
- At the same time, the elimination of these problems largely depends on the decision of the priority, strategically defined tasks facing the RF foreign energy policy, among which:
- diversification of export energy markets and commodity structure of exports;
- strengthening the positions of leading Russian energy companies abroad;
- reflection of Russia's national interests in the system of functioning of the world energy markets that ensures their predictability and stable development;
- ensuring effective international cooperation in relation to risky and complex projects in Russia;
- ensuring stable conditions in the energy markets, including demand security and reasonable prices for the main products of Russian energy exports.

5 Innovative Activity Among the Factors Determining the Development and the Need for the Unification of the Legal Regulation of Energy Relations

According to the Energy Strategy of Russia for the period until 2030, approved by the Order of the Government of the Russian Federation of November 13, 2009 No. 1715-r (Order of the Government of the Russian Federation 2009), the solution of the above tasks "is carried out with the use of diplomatic support of the interests of Russian fuel and energy companies abroad, as well as by applying some measures and mechanisms of the state energy policy, such as:

- active participation in the international negotiation process on energy issues, ensuring a balance of interests of importers, exporters and transiters of energy resources in international treaties and the activities of international organizations;
- Development of cooperation in the field of energy with the countries of the Commonwealth of Independent States, the Eurasian Economic Community, Northeast Asia, the Shanghai Cooperation Organization, the European Union, with other international organizations and states;
- development of new forms of international (including technological) cooperation in the energy sector;
- Russia's active participation in international cooperation to develop the energy of the future (hydrogen energy, thermonuclear energy, use of energy from sea tides, etc.);
- coordination of activities in the world oil and gas markets with the countries of the Organization of Petroleum Exporting Countries and the Forum of Gas Exporting Countries;
- assistance in the formation of a unified European-Russian-Asian energy space;
- assistance in ensuring a favorable and non-discriminatory regime for the activities of domestic energy and service companies (as well as foreign companies with Russian participation) in world markets, including their access to foreign energy markets and final energy markets;
- assistance in attracting foreign investments on mutually beneficial terms, primarily in technically complex and risky projects;
- ensuring the access of Russian energy companies to the use of resources of the world financial markets, advanced energy technologies."

Thus, today the process of formation of international markets for energy resources with the participation of the Russian Federation can't ignore such objective factors of the development of social relations in the area under investigation, influencing their legal regulation as the need for international legal unification of the legislation of member states of international associations with the participation of the Russian Federation or its main foreign economic partners. Paying attention to the special significance of the unification of the legal regulation of the foreign trade turnover of energy resources within the framework of integration processes and the creation of a single legal space for such intergovernmental associations as the EU, the CIS, the EAEC, the BRICS, a separate chapter of the monographic research is devoted to the study of these issues.

In addition, an analysis of the above problems, characteristic of the current foreign trade turnover of Russia's energy resources, as well as the tasks and activities identified in the national strategic acts for the strategic development of the country's fuel and energy complex, allows to draw a conclusion that the determining factor is also the level of innovative activity in the energy sector.

Indeed, a characteristic feature of modern world economic development as a whole is the transition of leading countries to building an economy, the foundation of which is the creation, dissemination and use of knowledge. All the developed countries of the world are involved in the process of developing and implementing measures aimed at innovative development and comprehensive support of innovative entrepreneurial activities. The sphere of energy is not an exception.

In the framework of traditional business ideas typical for foreign entrepreneurs, innovation is a commercial activity connected with the receipt of a new product and its sale to other market participants (Gribanov 2016; Lammers and Diestelmeier 2017).

Innovative activity includes all scientific, organizational, technological, commercial and financial steps that, in fact or intentionally, lead to the implementation of innovations—the implementation of a new or improved product (product or service) introduced into the market, or the technological process used in practice (Inshakova and Ryzhenkov 2015).

Innovative activity is an effective tool for commercializing the achievements of scientific and technological progress, becoming the defining element of the international competitiveness of any country.

Innovative activity in the field of energy gives hope for resolving one of the priority problems of the 21st century—the depletion of energy resources. Russia still occupies the place of a "raw donor" of the Western world and developing economies of the Asian region, which clearly does not correspond to the potential and functions of the Russian state on the world stage.

6 State Policy of Development of the Energy Sector and Innovations

Chairman of the Government of the Russian Federation Medvedev December 22, 2016 at a meeting on the draft Energy Strategy of Russia for the period until 2035 said: "If we want to have competitive energy in the future, then the state and business need to invest in innovations" (Government of Russia 2017).

The Concept of long-term social and economic development of the Russian Federation for the period until 2020 is approved by the Order of the Government of the Russian Federation of November 17, 2008 No. 1662-r (Order of the

Government of the Russian Federation 2008a), one of the tasks of which is the creation of a competitive economy of knowledge and high technologies.

The transition of the Russian economy from the export-raw to the innovative socially-oriented type of development will significantly expand the competitive potential of the national economy by increasing its comparative advantages, in science, education and high technologies, and on this basis new sources of economic growth and human welfare will be used (Lavrijssen 2016).

One of the directions of transition to an innovative socially oriented type of economic development is the consolidation and expansion of Russia's global competitive advantages in traditional spheres, such as:

- expansion of stable supplies of energy resources to the largest consumers of the world market, product and geographical diversification of energy exports;
- the transition from the export of primary energy resources to the export of their deep processing products (Kostinboy 2016);
- development of major nodes of the international energy infrastructure in the Russian Federation, which use new energy technologies;
- the achievement of the leading positions in the development of renewable energy sources and the introduction on an industrial scale of environmentally friendly energy production technologies.

Directive of the Government of the Russian Federation No. 2227-r of 08.12.2011 approved the Strategy for Innovative Development of the Russian Federation for the period up to 2020 (Order of the Government of the Russian Federation 2008b), a strategy that defines the goals, priorities and instruments of state innovation policy and long-term development guidelines for innovation actors, as well as the benchmarks for financing the sector of basic and applied science and supporting the commercialization of developments.

Section 4 of the Strategy notes that ensuring progress in the transition of the economy to an innovative development path is impossible without increasing efficiency, reducing resource intensity and expanding the redistribution of energy and raw materials industries.

The construction of the country's intellectual energy infrastructure, the creation of a national energy sector of a new technological level, the formation of scientific and technical reserves exceeding the world level is the future result of the most developed national economies.

Significant importance is achieved by saving resources, introducing the best available highly effective domestic technologies, increasing the environmental friendliness of production and the level of processing raw materials, switching to modern types of raw materials and fuel, as well as the development of energy, based on the use of alternative and renewable energy sources, which should become an important factor in the innovative development of the economy. Innovative programs of the largest public sector companies, technology platforms in the energy sector and cooperation with leading international companies will be an important tool for solving the problems of the modernization of the commodity sectors.

The basic principles of energy development are fixed in the Energy Strategy of Russia for the period until 2030, approved by the Order of the Government of the Russian Federation of November 13, 2009 No. 1715-r (Federal Law 2003) (*hereinafter—the Strategy*).

The main directions of the long-term development of the raw materials industries, as envisaged by the Energy Strategy of Russia for the period up to 2030, are: integration into the world energy system; change in the structure and scale of energy production; transition to the path of innovative development and increase of energy efficiency; creation of a competitive market environment.

The creation of an innovative and efficient national energy sector of the country that is adequate to the needs of a growing economy in energy resources and Russia's foreign economic interests is the main target of the Strategy.

The action plan ("road map") "Introduction of innovative technologies and modern materials in the fuel and energy sector" for the period until 2018 is approved by Decree of the Government of the Russian Federation No. 1217-r dated July 3, 2014 (Order of the Government of the Russian Federation 2014) (*hereinafter—the Plan*).

The plan is a set of measures to improve the mechanisms of government incentives and support the introduction of innovative technologies and modern materials in the branches of the fuel and energy complex of the Russian Federation by creating a unified system for managing innovation activities in the fuel and energy sector.

The plan's activities are aimed at providing the Russian fuel and energy complex with innovative and scientific and technical solutions in the amount necessary to maintain Russia's energy security, the relevant regulatory and legal framework of documents, as well as highly efficient equipment and technologies (Kostinboy 2016). The implementation of the activities of the plan helps to overcome existing barriers, develop innovative infrastructure in the fuel and energy sector, create favorable conditions for the development of innovative activities of economic entities.

The plan notes that the creation at the federal level of a unified system for managing innovation activities in the fuel and energy sector will ensure the effective and comprehensive implementation of mechanisms to support priority sectoral projects, improvement of regulatory and legal regulation and coordination of interaction of relevant instruments (innovation development programs, development institutions, business associations, technological platforms and innovative territorial clusters).

Support for innovative research and development projects in the energy sector is carried out within the federal target program "Research and development in priority areas of the development of the scientific and technological complex of Russia for 2014–2020", approved by Ordinance of the Government of the Russian Federation of 21.05.2013 No. 426.

In November 2016, the Minister of Energy of the Russian Federation, Alexander Novak, approved the Forecast of the scientific and technological development of the fuel and energy sector of Russia for the period up to 2035 (*hereinafter—the*

Forecast of the STD)—the basic document of scientific and technological development in the fuel and energy complex. It defines promising areas of research and development of technologies, as well as sets targets for industry participants in the development and implementation of innovative technologies and advanced materials in the fuel and energy complex (Ministry of Energy of the Russian Federation 2017a).

Taking into account the Forecast of Scientific and Technical Development of the Ministry of Energy of Russia, a draft of the Energy Strategy of Russia for the period up to 2035 (Ministry of Energy of the Russian Federation 2017b), a plan that takes into account the results of the annual monitoring of the implementation of the Energy Strategy of Russia for the period up to 2030.

One of the priorities of the state energy policy for the development of the energy sector is the stimulation and support of innovative activities of the organizations of the fuel and energy complex and related industries in the direction of increasing the efficiency of the use of fuel and energy resources and the production potential of the entire fuel and energy complex.

The Strategy provides for the solution of two main tasks for an effective response to the challenges of technological development and overcoming the accumulated problems of innovation in the fuel and energy sector: development of the segment of the national innovation system in the fuel and energy complex and the development of network forms of organization and promotion of innovation and technology transfer, including the expanded integration of Russian networks into international networks.

7 Development of Mechanisms for Legal Support of Innovative Activities in the Energy Sector

In the opinion of V.V. Romanova, with whom we agree, further development of legal support for innovation activity in the energy sector should be carried out on condition of close interaction of science, the state, energy companies with coverage of the main links of the innovation chain, in the links of which: formation of an order for an innovative project and identification of sources of financing; implementation of an innovative project; registration of rights to the results of intellectual activity; realization of the results of intellectual activity; protection of rights to the results of intellectual activity. Essential assistance in identifying problem aspects of the gaps in the legal regulation of innovation activities in the energy sector can be provided by sectoral research, expert, educational centers created on the basis of scientific and higher educational institutions in cooperation with energy companies.

Taking into account the state innovation policy in the energy sector, reflected in the above-mentioned strategies, state companies develop and implement their innovative development programs—the main tools for achieving innovative development.

In accordance with the Russian Federation Government Ordinance No. 373-r of 06.03.2015 "On approval of the plan for the implementation in 2015—2016 of the Strategy for Innovative Development of the Russian Federation for the period to 2020" (Order of the Government of the Russian Federation 2011) joint-stock companies with state participation, state corporations and federal state unitary enterprises are instructed to update the innovative development programs aimed at increasing the output and consumption of innovative products and increasing labor productivity, and ensure their effective implementation.

8 Case Study

In the passport of the Innovative Development Program of PJSC Gazprom until 2025 (Gazprom 2017), it is noted that the development of PJSC Gazprom as a global energy company and a reliable energy supplier is associated with the permanent solution of various tasks that require the search, acquisition and application of new knowledge, increase of activity and efficiency of innovative activity.

This Program contains a set of interrelated activities aimed at developing and introducing new technologies, innovative products and services that correspond to the world level, as well as creating favorable conditions for the development of innovative activities both in PJSC Gazprom and in adjacent areas of industrial production in Russia.

The most important part of the development concept of PJSC NK Rosneft is the Program of its innovative development. Innovative activity is aimed at creating and implementing new technologies for solving production problems, as well as modernizing the industrial base.

The company faces global challenges: increasing the oil recovery factor and the depth of its processing, replenishing reserves, maximizing the full use of associated petroleum gas, effective implementation of offshore projects, minimizing capital and operating costs, improving energy efficiency, as well as environmental and industrial safety. The innovative development program of PJSC NK Rosneft is called upon to answer the challenges facing it and ensure the achievement of its strategic targets (Rosneft 2017).

Prospective directions of innovative development for the long-term period are formulated in the Passport of the Innovative Development and Technological Modernization Program of the State Corporation Rosatom for the period up to 2030 (in the civil part) (Rosatom 2017).

With the aim of introducing advanced developments in the field of exploration and production in the PJSC "LUKOIL", an Innovation Cooperation Portal—an electronic platform in which any supplier of effective technologies and equipment

—from large enterprises and industry institutes to small organizations and individuals—can offer its work for use in the company (Lukoil 2017).

9 Conclusion

Thus, it is obvious that the level of innovative development in the energy sector is becoming one of the leading factors affecting the efficiency of the foreign trade turnover of energy resources.

Innovative activity in the field of energy is manifested, first of all, in the implementation of the latest high-performance technologies and equipment, in the use of advanced world experience, increasing the environmental friendliness of production and the level of processing of raw materials, the transition to modern types of raw materials and fuels, as well as the development of energy, based on the use of alternative and renewable sources of energy.

We believe that further development of the legal content of strategic planning documents in the energy sector should be carried out taking into account the tasks envisaged in the relevant programs of energy companies, the needs of energy companies of various energy sectors in the development of innovative technologies (Romanov 2016).

Summarizing the doctrinal opinions, as well as the official data on priority tasks and a set of priority actions of innovative development of the largest Russian companies—suppliers of energy resources we can conclude that in the list of main tasks of development of legal regulation in the sphere of foreign trade turnover of energy resources—creation of a legal basis for the transition of the Russian economy from the export-raw material to the innovative socially-oriented type of development. With the aim of forming a "new energy" based on renewable energy sources, distributed generation, intellectualization and the creation of a competitive market environment. The solution of this task is connected with a complex of legal measures, which should be reflected not only in the framework documents of a program nature. Among them are the Energy Strategy of Russia for the period up to 2030, the Concept of Long-Term Social and Economic Development of the Russian Federation for the Period to 2020, the Strategy for Innovative Development of the Russian Federation for the Period to 2020, Plan of measures ("road map") "Introduction of innovative technologies and modern materials in the fuel and energy sector" for the period until 2018, Forecast of scientific and technological development of the fuel and energy sector of Russia for the period up to 2035. Their provisions should be specified in unified international legal instruments and domestic legislation. Because the available, analyzed in this chapter array of policy documents in the field of innovative energy development, including the turnover of its resources, has a recommendatory-declarative character and therefore is not able to introduce economic entities in a strictly defined framework and have a significant impact on the legislative process.

References

Federal Law. (2003). *On the Basics of State Regulation of Foreign Trade Activity, (accepted the State Duma of November 21, 2003 no. 164-FL)*. (Russian).

Frolov, D. P. (2017). Variation field of transplantation of the European institutions of innovative development. *Journal of Economic Theory, 1*, 116–132.

Gazprom. (2017). *List of insider information*. Retrieved January 26, 2017, from http://www.gazprom.ru/f/posts/97/653302/prir-passport-2016-11.pdf. (Russian).

Government of Russia. (2017). Meeting on the draft Energy Strategy of Russia for the period until 2035. Retrieved January 25, 2017, from http://government.ru/news/25812/.

Gribanov, D. V. (2016). Problems of the legal regulation of innovation in national and international legislation. *Business, Management and Law, 1–2*, 29–36.

Inshakova, A. O., & Ryzhenkov, A. Y. (2015). *Institutional analysis of the nano technological «Revolution»: Synthesis of economics and law*. St. Petersburg: Aleteya.

Kostinboy, A. S. (2016). *The Organizational-economic mechanism of realization of a regional power policy*, Pskov.

Lammers, I., & Diestelmeier, L. (2017). Experimenting with law and governance for decentralized electricity systems: Adjusting regulation to reality? *Sustainability* (Switzerland), *9*(2), 212.

Lavrijssen, S. (2016). The right to participation for consumers in the energy transition. *European Energy and Environmental Law Review, 25*(5), 152–171.

Lukoil. (2017). Innovative cooperation. Retrieved January 26, 2017, from http://www.lukoil.ru/Business/technology-and-innovation/InnovativePartnership. (Russian).

Ministry of Energy of the Russian Federation. (2017a). *The forecast of scientific and technological development of the fuel and energy sector of Russia for the period up to 2035*. Retrieved January 26, 2017, from http://minenergo.gov.ru/node/6365. (Russian).

Ministry of Energy of the Russian Federation. (2017b). The draft energy strategy of the Russian Federation for the period up to 2035. Retrieved January 26, 2017, from http://minenergo.gov.ru/node/1920. (Russian).

Order of the Government of the Russian Federation. (2008a). *Concept of long-term social and economic development of the Russian Federation for the period until 2020, (approved by decree of the Government Russian Federation of November 17, 2008, no.1662-r)*. (Russian).

Order of the Government of the Russian Federation. (2008b). *Strategy of innovative development of the Russian Federation for the period until 2020, (approved by the Decree of the Government of the Russian Federation of December 8, 2011 no. 2227-r)*. (Russian).

Order of the Government of the Russian Federation. (2009). *On the Energy Strategy of Russia for the period until 2030, (approved by the decree of the Government of the Russian Federation of November 13, 2009 no. 1715-r)*. (Russian).

Order of the Government of the Russian Federation. (2011). *On the approval of the plan for the implementation in 2015 – 2016 of the Strategy for Innovative Development of the Russian Federation for the Period to 2020, (approved by the decree of the Government of December 8, 2011 no. 2227-p)*. (Russian).

Order of the Government of the Russian Federation. (2014). *«Introduction of innovative technologies and modern materials in the fuel and energy sector» for the period until 2018, (approved by the order of the Government of the Russian Federation of July 3, 2014 no.1217-r)*. (Russian).

Romanov, V. V. (2016). Legal support of innovative activity in the sphere of power engineering. *The legal World, 7,* 43–48.

Rosatom. (2017). *Passport of the program of innovative development and technological modernization of the State Corporation Rosatom for the period up to 2030 (in the civil part).* Retrieved January 26, 2017, from http://www.innov-rosatom.ru/files/articles/b5477a6c913a8eba1ed55ebb453250c1.pdf. (Russian).

Rosneft. (2017). *Science and innovation.* Retrieved January 26, 2017, from https://www.rosneft.ru/Development/sci_and_innov/. (Russian).

Innovative Technologies of Oil Production: Tasks of Legal Regulation of Management and Taxation

Agnessa O. Inshakova, Alexander I. Goncharov and Elena I. Inshakova

1 Annotation

This chapter of the monograph substantiates the need for intellectualization of the main factors of production in the oil industry at the present stage of development and modernization of the Russian economy The direct dependence of the financial security of the Russian Federation on the volume of extracted raw materials, the severity of issues related to the delineation of powers in the sphere of owning, using and disposing of natural oil and gas resources is considered in the chapter in the context of the course announced by the authorities for innovation and high technologies. Introduction of innovative technologies in the field of hydrocarbon production is considered as a priority component of the general direction within the framework of modernization of the Russian economy The relevant tasks of the Russian oil industry, consisting of large vertically integrated oil companies, are being studied. The ways of finding acceptable, economically adequate legal bases for effective mining of mineral resources are discussed. Which are in direct dependence on innovative technologies that satisfy the social, economic and political aspirations of the state, and allow to solve the problem of limited natural resources. An innovative approach to taxation of oil companies is being explored, providing for the withdrawal in favor of the state of half of the net income actually received by the economic entity, after covering all the costs of oil production.

A. O. Inshakova (✉) · A. I. Goncharov
Department of Civil and International Private Law, Institute of Law, Volgograd State University, Volgograd, Russia
e-mail: gimchp@volsu.ru; ainshakova@list.ru

A. I. Goncharov
e-mail: gimchp@volsu.ru; goncharova.sofia@gmail.com

E. I. Inshakova
Department of Economic Theory, World and Regional Economics, Institute of Economics and Finance, Volgograd State University, Volgograd, Russia
e-mail: inshakovaei@volsu.ru

© Springer International Publishing AG, part of Springer Nature 2019   79
O. V. Inshakov et al. (eds.), *Energy Sector: A Systemic Analysis of Economy, Foreign Trade and Legal Regulations*, Lecture Notes in Networks and Systems 44, https://doi.org/10.1007/978-3-319-90966-0_6

2 Materials

The tasks of legal regulation of management and taxation in oil production with the increasing use of innovative technologies in this chapter of the monograph are solved on the basis of a complex of sources. United Nations documents used the United Nations Convention on the Law of the Sea (UNCLOS) (concluded in Montego Bay on 10.12.1982) (as amended on 23.07.1994). The normative acts that are part of the system of Russian law and regulate relations in the field of extracting energy resources in the fields, as well as the relationships of subsequent taxation of this business, were studied. The provisions of the federal laws in the sphere of energy regulating relations in the fuel and energy complex are analyzed, these include: Federal Law No. 256-FZ of July 21, 2011 "On the Safety of the Fuel and Energy Complex"; Law of the Russian Federation No. 2395-1 of February 21, 1992 "On Subsoil"; The Tax Code of the Russian Federation (part two) of 05.08.2000 No. 117-FZ (as amended on November 27, 2017); Federal Law No. 189-FZ of July 11, 2011 "On Ratification of the Agreement on the Organization, Management, Functioning and Development of Common Oil and Oil Products Markets of the Republic of Belarus, the Republic of Kazakhstan and the Russian Federation"; Federal Law of July 21, 1997, No. 122-FZ "On state registration of rights to real estate and transactions with it." Studied GOST R 52104-2003. National standard of the Russian Federation. Resource-saving. Terms and definitions (approved by the Decree of the State Standard of the Russian Federation of July 3, 2003 No. 235-st) (as amended on November 30, 2010). The draft federal laws "On Amendments to the Tax Code of the Russian Federation (regarding the introduction of a tax on additional income from the extraction of hydrocarbon raw materials)" were studied (Government Order No. 2613-r of November 25, 2017) and "On Amendments to Articles 31 and 35 of the Law of the Russian Federation" On Customs Tariff "(Order of the Government of the Russian Federation No. 2614-r of November 25, 2017)—prepared by the Ministry of Finance of Russia in accordance with the decisions on the results of the meeting of the Government Commission on the development of the fuel and energy complex and improving the energy efficiency of the economy on June 14, 2017. Along with the documents related to legal regulation, the provisions of the National Project "Creation of a complex of domestic technologies and high-tech equipment for the development of reserves of the Bazhenov's svita" were considered.

The scientific base of this chapter of the monograph is formed on the basis of research works of Russian scientists. In developing the issues of classification, reserves, dynamics and taxation of the extraction of Russian natural resources, the authors attracted the scientific works of a number of researchers, namely: Dovbnya, Perchik, Popondopulo, Gorodov, Petrov, Muslimov, Petrova, Salieva, Tokarev. A study of the problems of innovation in oil production for rocks of the shale formation, approaches to taxation for new deposits predetermined the authors' appeal to the scientific works of Professor A.O. Inshakova, to the published information materials of the companies: GazpromNeft, TASS. Discussion questions of optimization of taxation for new deposits and stimulation of investments in the Arctic were considered by the authors on the basis of empirical data of VYGON Consulting, and also scientific works of a number of scientists, such as: Motin, Trofimov, Pevzner. In preparing the conclusion of the monograph chapter, the scientific positions of a number of Russian scientists, including: Ageev, Kalinin.

3 Methods

The scientific development of the content of this chapter of the monograph was carried out on the basis of the universal scientific method of historical materialism. General scientific methods of cognition are used: dialectical, hypothetic-deductive method, generalization, induction and deduction, analysis and synthesis, empirical description, classification. The research also used private-science methods: legal-dogmatic, method of concrete sociological research, comparative legal analysis, mathematical method, structural-functional method, etc.

4 Introduction

Intellectualization of the main factors of production as the basis for the formation of the national innovation economy is characteristic, first of all, for such a branch of management as extraction, production and processing of oil. Indeed, speaking of innovations in the context of existing reality, from a pragmatic point of view, considering their applied characteristics for the economic turnover of the country, first of all, one remembers not the start-ups of Skolkovo, but the branches of the national economy that ensure the filling of the state budget. The special significance of involving in the national economic turnover the results of scientific and scientific and technical activities in the oil sector is due to the worldwide recognition of Russia as an important participant in the energy market. Despite the announcement by the authorities of the country of a course on innovation, high technologies, including nanotechnology, the Russian state is still directly dependent on the volume of extracted raw materials, the degree of its processing, and the possibilities of transporting oil and gas products. Also, various issues related to the delineation of powers in the possession, use and disposition of natural oil and gas resources are required to be resolved (Dovbnya 2012). We believe that today the priority issue in the mainstream of the modernization of the Russian economy is the introduction of innovative technologies into the sphere of hydrocarbon production. Russia provides 12% of world oil trade. The length of the main oil pipelines of companies of Russian nationality is about 50 thousand km, oil pipelines—19, 3 thousand km (Official website of the Ministry of Energy of the Russian Federation 2017). The strategic importance of the above parameters for the countries-buyers becomes obvious on the example of the European Union, 25% of the import of oil and gas to which comes from Russia. So, over 80% of the volume of Russian oil is exported to European countries, Russia's share in the markets is about 30%.

The oil industry of Russia includes oil producing companies, oil refineries and companies for the transportation and marketing of oil and petroleum products. The structure of the oil industry includes 10 large vertically integrated oil companies. Among the others are Rosneft, Lukoil, Surgutneftegaz, TNK-BP, Gazpromneft and others (Official website of the Ministry of Energy of the Russian Federation 2017).

Therefore, the Russian Federation is in continuous search for an acceptable legal basis for effective mining (Shakhverdiev Dutov and Shakhverdiev 2009). It should also be borne in mind that such legal frameworks are now directly dependent on the innovative technologies being implemented, which the oil industry badly needs (Inshakova 2014). In addition, Russian oil-producing companies are actively interested in actively introducing innovations in their business, first of all, due to the awareness of the limited natural resources.

5 Classification, Reserves, Dynamics and Taxation of Extraction of Russian Natural Resources

As you know, all natural resources are divided into two fundamentally different types:

- renewable resources, which include wind energy, solar, water, plant and other energy;
- non-renewable resources—contained in the subsoil (oil, gas, coal and others) (Perchik 2006).

The definition of non-renewable resources in the most general form is contained in the National Standard "Resource Saving". Terms and definitions "(GOSTR 52104-2003" 2003). According to paragraph 4.16, non-renewable resources are part of the natural resources involved in economic activities that are converted into products and converted to waste at the stages of the product life cycle. For example, they include oil, gas, coal, peat, shale and others. Based on this general definition, experts define non-renewable energy resources as part of the natural resources involved in economic activities that are converted into energy (Salieva 2013). In addition, researchers point to the inability to provide an exhaustive list of non-renewable energy resources or non-renewable energy sources (Popondopulo et al. 2011), since in connection with the expansion of knowledge about the environment, new primary energy sources are involved in economic circulation (Muslimov 2009). Among them, for example, oil shale.

Reducing the stocks of light oil and medium-density oil—traditional oil, puts oil companies in the task of developing alternative sources of hydrocarbons, implemented through innovative technologies. One of the priority areas, both in the opinion of scientists and practitioners—representatives of the largest Russian companies in the field of oil production, is the development of oil shale. The media also firmly established the concept of "Shale revolution". Shale oil is oil that is extracted from shale spits or from other rocks with very low permeability by drilling horizontal wells and multi-stage fracturing (Petrova 2010).

In February 2013, Price Water House Coopers (PWC) (PwC: Audit and assurance, consulting and tax services 2017) prepared the Forecast for the Development of the Shale Oil Market, according to which, by 2035, the production of shale oil in the world will grow and accordingly, oil prices will fall by 25–40%. As a result, the overall gain of the world economy from the production of cheap and affordable oil is obvious. However, it is time for the national expert community to take timely measures to prevent the loss of huge revenues by the Russian state.

The objectives of this study will be assisted by a comparative legal analysis with legislative regulation of the studied issues of the United States, where the advanced innovative technologies of multi-stage hydraulic fracturing of reservoirs, which have provided a significant increase in oil and gas production, are already being used to the full.

According to the PWC report in 2004, 111000 barrels of oil were produced in the US in oil shale per day (b/d), and in 2011—already 553000 b/d (growth was 26% per year). U.S. production of tight oil has increased significantly since 2010, due to the technological improvements that have resulted in the drilling costs reduction and drilling efficiency improvement in major shale basins such as the Permian Basin, the Bakken, and the Eagle Ford, etc. Tight oil production exceeded 50% of total U.S. oil production in 2015 when it reached 4.9 million b/d (Tight oil expected to make 2018).

As to the U.S. Energy Information Administration forecast, shale oil production will grow by 111,000 barrels per day (b/d) to 6.55 million b/d in February, 2018 (SNBC 2018).

In this situation, United States of America could become the world oil leader in 2018. The U.S. hasn't got this global status, not ahead of both Russia and Saudi Arabia, since 1975 (CNN Money 2018).

Volume of tight oil production, based on the optimistic technology and resource scenario of the U.S. energy sphere development, will reach 11.0 million b/d by 2035, or 66% of total U.S. production. In case of the pessimistic scenario realization, tight oil will provide less than 50% of total oil production after 2030, and in 2040 total U.S. oil production will be below its current level (Tight oil expected to make 2018).

At present, US oil imports have fallen to a minimum value for 25 years. The US Energy Information Administration forecasts that the production of shale oil will replace 35–40% of the sea oil imports in the US (U.S. Energy Information Administration (EIA) 2017). This oil will go to other countries, first of all, to China.

Such a rapid rate of increase in domestic production of shale oil in the US has already put pressure on prices downward. According to PWC experts, world oil prices may soon fall even below these forecasts.

Successful production of shale oil in the US can be an example of the global development of technologies for the production of shale oil for other countries in the world where shale oil production is practically not being conducted (Grinets 2014). It is obvious that the consequence of PWC's projected decline in oil prices will be the numerous losses of producers and exporters of traditional types of oil, the list of which headed by Russia by a large-scale criterion.

Therefore, the results of the assessment of the long-term development of shale oil production in the world are impressive for PWC experts, and the possible consequences for the Russian Federation are alarming and make scientists, including economic and legal scientists, use their research potential.

Due to the fact that at the current stage the country's budget policy is directly dependent on oil and gas revenues, we believe that price is one of the most important risks.

Even in the case of a small reduction in oil prices, serious negative consequences can occur, taking into account the example of 2012, when the budget became deficit, provided that the price of oil was 117.2 dollars per barrel. And also, taking into account that in December 2017 the price of oil was 63.2 USD per barrel.

The biggest Russian oil companies, which determine the image of the state as the largest exporter of traditional oil, are also concerned about this price situation by attempting to overproduce and introduce innovative technologies in the development of shale oil production. Moreover, it should be noted that the resource potential of shale oil in Russia is huge, which is another argument in favor of paying close expert attention to the problem now that traditional reserves of prospected oil sources are increasingly difficult to extract. Indeed, deposits of oil shale throughout Russia, especially in Western Siberia, where such low-permeable layers saturated with oil are called svities (for example, the Bazhenov's svita in which, according to the experts' estimations, the world's largest reserves of shale oil are located) can be considered as the most promising alternative source of traditional oil.

Why is the development of shale oil production successfully conducted by US oil companies in the Russian Federation considered irrational? Experts call inefficient taxation in the list of problems (in addition to the technical difficulties of their development due to the different geological shale structure in Western Siberia, which makes it impossible to literally transfer foreign technologies) that hamper the development of domestic shale deposits.

This problem becomes especially urgent due to the necessity of using tax incentives to intensify the efforts of leading Russian oil companies to develop domestic technologies for the industrial production of shale (as well as offshore) oil, taking into account the limited possibility of attracting the largest foreign corporations under the imposed sanctions.

The extraction of all minerals is taxed accordingly. The question of reforming the taxation system of the oil industry is very acute, by introducing a special tax on the added income.

These types of taxes have a fundamental difference. According to Article 261 of the Tax Code of the Russian Federation, absolutely all volumes of mined resources are subject to a mineral extraction tax, in which the costs of the extracting enterprise at the stage of development of natural resources act as expenses, reducing the tax base for the profit tax of the organization. In turn, there is the possibility of taxation of net income arising from the extraction of natural resources, which is called the tax on added income, in which only the difference between the cost of sold oil and the cost of its extraction is taxed. The use of such a tax would encourage more active investment in the development of new deposits. In such a situation, the tax is not levied until the full payback of capital expenditures.

The need for the introduction of the tax on the added income in the sphere in question is spoken by both theoreticians and practitioners of the oil industry. In particular, the president of OAO "LUKOIL" V. Yu. Alekperov argues for this necessity, first of all, by the fact that at present the extraction of oil from existing fields is at the final stage, while new fields have not been developed yet (Vedomosti 2013). It is necessary to activate geological exploration in order to develop them, including small and medium-sized deposits. The introduction of the tax on added income is intended to contribute to these activities.

The concept of the tax on added income was supported by the academic community (Tokarev 2013). Specialists believe that the introduction of a tax on added income is particularly suitable for new projects with a proportionate reduction in the tax rate on extraction. One can't but agree that the introduction of such a tax will make it possible to master practically any deposits of a high-cost continental shelf, including shale oil. At the same time, according to estimates of specialists, the state's revenues, in comparison with the current tax legislation, will amount to 90%. According to the publishing house "Rossiyskaya Gazeta" in January 2013, the company began drilling the first well to study the Bazheno-Abalakskoye horizon of the Krasno-Leninsky deposit.

6 Case Study

There are notable advances in the production of shale oil in Russia. Subsidiary of PJSC Gazprom—Gazpromneft in 2012 launched a pilot project to develop the Bazhenov's svita of the Verkhne-Salym field in Ugra.

According to the publishing house "Rossiyskaya Gazeta" in January 2013, the company began drilling the first well to study the Bazheno-Abalakskoye horizon of the Krasno-Leninsky deposit. In August 2016, at the Palyanovskaya area of the Krasnoleninskoye field, Gazpromneft-Khantos completed the construction of a well with a horizontal section of 1 thousand meters to develop unconventional reserves —the Bazhenov's svita.

In a high-tech well, a 9-stage hydraulic fracturing of the formation was carried out, a gushing flow of anhydrous oil from a productive horizon located at a depth of 2.3 thousand meters was obtained. The debit is more than 45 tons of oil per day. As part of a project to develop unconventional reserves in Yugra, Gazpromneft was the first company in Russia to implement the entire cycle of technological solutions used in the global oil and gas production to develop shale oil. In particular, the horizontal section of the well was cemented with elastic cement, followed by a multi-stage hydraulic fracturing of the formation with high injection rates of the process fluid. Elastic cement, in contrast to conventional, is resistant to repeated exposure to variable loads during the multi-stage hydraulic fracturing of the formation and allows reliable insulation of the created cracks from each other. The combination of reliable insulation and high injection rates during hydraulic fracturing makes it possible to create an intensive network of cracks along the entire length of the horizontal trunk, thereby increasing the volume of hydrocarbons involved in the development (Neftegaz 2017).

7 Innovations in Oil Production for Rocks of the Shale Formation, Taxation for New Oilfields

In May 2017, the Ministry of Energy of the Russian Federation assigned the status of a national project "Creation of a complex of domestic technologies and high-tech equipment for developing reserves of the Bazhenov's svita". Within the framework of the national project, it is planned to develop and implement technologies for prospecting oil and gas bearing Bazhenov's deposits based on domestic methods of seismic, magnetic, gravity and specialized core research, as well as geological and hydrodynamic modeling of reservoirs. It is also planned to create Russian technologies for the construction of horizontal wells with multi-stage hydraulic fracturing of the reservoir, optimized for mining and geological conditions of the Bazhenov's svita, to create ways to involve light oil from the rocks of the shale formation in development of reserves due to thermochemical methods of influence. In case of successful implementation of all the technologies created, according to the strategy approved by Gazprom-Neft for the development of a resource base for unconventional sources of hydrocarbons, the target level of production from the deposits of the Bazhenov's svita in 2025 may be about 2.5 million tons per year. The main research work under the national project will be carried out until 2021. 2022–2025 years will be a period of industrial introduction of technologies and their replication to the domestic and foreign markets (Gazprom Neft 2017).

During 2017, the government discussed the reform of taxation of the country's oil industry on the basis of introducing a tax on additional income from the extraction of hydrocarbon raw materials for pilot projects and the complete abolition of customs export duties. At the end of November 2017, the Government of Russia submitted to the State Duma a draft law on a tax on additional income in the oil industry. In the current period, oil companies pay mineral extraction taxes and export duties, these amounts are withheld from revenue. The tax on additional income should partially replace the tax on extraction of minerals. The rate of the tax on the additional income will make 50%, but unlike the tax to extraction of minerals it will be charged not with volumes of extracted oil. The basis for calculating the tax on additional income will be the estimated revenue from operating and investment activities for exploration and production, reduced by the amount of actual expenses, by the amount of the tax on extraction of minerals and transportation costs. Deputies of the State Duma must adopt the law in the first quarter of 2018, the law will enter into force on January 1, 2019.

The tax on additional income is planned to be extended to four groups of deposits. The first group—new fields in Eastern Siberia with a depletion of less than 5%; the second group—deposits that enjoy a benefit on export duty; the third group —operating deposits in Western Siberia with depletion from 10% to 80% (with a quota for extraction of not more than 15 million tonnes); the fourth group is new deposits in Western Siberia with a depletion of less than 5% with aggregate reserves of no more than 50 million tons per year.

For new deposits, the draft law provides for the following preferences: a grace period for the payment of a mineral extraction tax, the transfer of losses of the current period to future periods, and accounting for losses of previous years for new deposits in new oil producing regions.

In the opinion of the Prime Minister of Russia D.A. Medvedev, the tax on additional income in the oil industry takes into account the project economy for the entire investment period, puts the tax burden, depending on the profitability of each field. The purpose of the new taxation system is to stimulate the development of new deposits and rational subsoil use. This system will be more flexible than the current taxation system, due to the dependence of the amount of taxes on the economic result of oil industry activities (TASS 2017).

The initiator of a new tax on additional income in the oil industry is the Ministry of Finance of the Russian Federation. The Ministry of Energy of the Russian Federation proposed, as an alternative, to impose a tax on the financial result for the mined fields with low-profit mining instead of the current mineral extraction tax. The tax on the financial result was supposed to be tested for 3–5 years in several pilot fields, after which it was decided whether to extend it to the whole industry.

The new tax mechanism is planned to be launched from 2018 on pilot development projects—the Ministry of Energy of the Russian Federation has identified 46 oilfields, 7 oil companies will participate in the pilot projects. In particular, applications for participation were filed by Gazprom Neft, Lukoil, Surgutneftegaz and Russneft. The total production of these deposits is about 7 million tons, they are located in Western Siberia. Rosneft has submitted to the Government of the Russian Federation a list of 15 fields with a total oil production of 6.14 million tons.

8 Discussion Issues of Tax Optimization for New Deposits and Promotion of Investments in the Arctic

Vygon Consulting research indicates that the transition to a tax on additional income from hydrocarbon production is not beneficial for new fields with low costs or with a significant share of the gas economy and for mature deposits with a high level of benefits. The tax base for the tax on additional income is calculated taking into account the income from sales of gas and gas condensate, and the tax on extraction of minerals for gas and gas condensate in the proposed model remains unchanged (VYGON. Consulting 2017).

The proposed bill on the tax on additional income from the extraction of hydrocarbon raw materials provides for a voluntary transition to it for two groups of pilot projects. The first group is new deposits in new regions, the second is mature fields in Western Siberia with total annual oil production for all fields of no more than 15 million tons. The current draft of the income tax bill has a balanced set of properties when applied to pilot projects, but it is not intended for the entire oil industry in Russia. The only way to apply the tax on additional income in the proposed parameters for the entire industry without losses is to maintain a mechanism for granting a tax privilege on the extraction of minerals of certain categories of stocks (VYGON. Consulting 2017).

According to Vygon Consulting experts, for most mature deposits without a tax on additional income, it is economically feasible for deposits with total extraction costs above \$5.5 per barrel. However in the presence of benefits the picture changes. For example, for super-viscous oil fields, the transition to an additional income tax is unprofitable under no circumstances. According to Vygon Consuting, the following fiscal mechanisms are known (VYGON. Consulting 2017):

- bonuses—one-time payments of the user of the subsoil to the state, for example a bonus at the time of obtaining a license and payment at the time the deposit was discovered;
- rentals—paid to the owner of the subsoil regularly for the right to use subsoil, are used mainly at the stage of prospecting and exploration;
- royalties—regular payment to the owner for the use of resources;
- extraction tax—regular payment; the rate is set either to the volume of extracted raw materials, or to its value;
- tax on random profits—excise tax; is established when the oil price exceeds a predetermined level;
- tax on excess profits—taxation of profits received in excess of the established "normal" (average) level;
- resource-rent taxes—taxation of net income.

Most oil-producing countries prefer mining taxes and others based on gross income, since the state thereby provides a certain level of financial revenues from the outset of production, in addition, it is easier to administer such taxes. Alternative examples of the advanced approach to taxation of the oil industry are the United Kingdom and Norway, as well as the United States, where systems based on the imposition of financial results are created. For example, the UK refused royalties, there are three taxes on oil production—corporate income tax, surcharge (on profits) and tax for "old" licenses. Norway also no longer uses royalties, corporate income tax and surcharge apply. In the US, an additional tax is not applied, there is a high profit tax and a significant role is played by bonuses, as a tool to receive payments at an early stage of the project.

It should be clarified that various benefits systems are used in world practice (Noguera 2017). This is a temporary or partial exemption from taxes for certain types of oil and gas fields: for deposits at an early stage of development; at a late stage of development; operated in extreme climatic and natural conditions; for deposits of high-viscosity oil, oil with a high sulfur content, hydrogen sulfide, paraffin and other impurities.

Along with optimizing the taxation of oil production, measures to stimulate foreign investment in the oil sector are of great importance, which would facilitate the acceleration of the development of hard-to-reach oil reserves and the intro-duction of innovations in relevant technological processes, characterized by high entrepreneurial risks and financial costs. Moreover, if the forecasts of the volume of shale oil reserves prove to be too high, then the question of developing the unpredictable Arctic with even higher environmental and economic risks will arise (Motin and Trofimov 2013). Obviously, the beginning of oil and gas production in the Arctic requires special risks to be taken into account at all stages of economic activity and additional costs. First of all, experts justify the need to create drilling platforms, tools and devices that can withstand the pressure of ice. Taking into account the climatic conditions of the North, as well as the physical and chemical properties of oil and gas, it can be argued that it is necessary to develop and introduce innovative technologies in hydrocarbon production processes, radically different from those used in warm countries. Thus, the researchers mention the increased consumption of fresh water, which is pumped into the bowels in the process of application and implementation of technology for the production of shale hydrocarbons in the Arctic (Motin and Trofimov 2013).

About their interests in the Arctic, said 17 countries, which makes us pay attention to the existence and quality of the relevant international legal regulation. Among countries, both located in this region, and just interested in extracting hydrocarbons on the Arctic shelf—Russia. Researchers warn of the threat of a legal conflict and territorial dispute due to the fact that in accordance with the United Nations Convention on the Law of the Sea of 1982, the shelf belongs to the country 200 miles from the coast (United Nations Convention 1982). And its borders can be extended for another 150 miles, if it is proved that the shelf is more extended.

As a number of transnational corporations participate in the plans for extracting minerals on the Arctic shelf, this circumstance will entail a clash of various foreign jurisdictions. In addition, assessing the risks of the relevant international legal vacuum, the researchers point out that "disputes over the Arctic have not been fully resolved to the present, and the estimated revenues from oil and gas production are quite high" (Motin and Trofimov 2013). As a result, appeals to international and national judicial bodies with various claims are expected. Consideration of claims takes quite a long time, therefore, the length of judicial procedures will lead to the fact that Russia's national interests may suffer.

In this context, the development of international cooperation of the Russian Federation with foreign partner countries through the conclusion of bilateral as well as multilateral treaties and agreements in the energy sector deserves the most positive assessment, as it has a positive impact on the transformation of the national mandatory energy regulation. The energy strategy of Russia for the period until 2030 defines as a strategic goal in the field of external energy policy the full integration of our country into the global energy market, strengthening its positions in this market and obtaining the greatest benefit for the national economy. The presence of Russia in the world energy market should have a legal anchorage corresponding to real modern needs.

9 Results

The significant role of legal mechanisms of international legal regulation of protection of private interests in bilateral agreements on the extraction of hydrocarbons is evident.

The basis for the implementation of the national strategic goal of the Russian Federation—full integration into the global energy market—is created by existing bilateral agreements and multilateral legal binding standards in the field of international energy relations.

On the active conclusion of agreements at the interstate level, which recently began to differ in more detailed elaboration of the main provisions, is evidenced by a number of documents recently ratified by the Russian Federation. For example, in September 2005, an Agreement on the organization, management, operation and development of a common oil and gas market was signed between the member states of the Eurasian Economic Community (Agreement 2005). The decision of the EurAsEC Interstate Council of December 9, 2010 No. 65 "On the course of implementation of the Action Plan for the formation of the Common Economic Space of the Republic of Belarus, the Republic of Kazakhstan and the Russian Federation" is to become the legal basis for the common oil and gas market of the member states of the Eurasian Economic Community. In July 2011, Russia ratified the Agreement on the Organization, Management, Functioning and Development of Common Oil and Oil Products Markets of the Republic of Belarus, the Republic of Kazakhstan and the Russian Federation (Federal Law 2011). Bilateral agreements on mutual cooperation in the oil sphere reached between Russia and Turkey (Protocol between the Government of the Russian Federation and the Government of the Republic of Turkey on Cooperation in the Oil Sphere (August 6, 2009, Ankara)), defined the vector of development of foreign economic cooperation by encouraging the creation of joint companies between Turkish and Russian oil companies in the Russian Federation, Turkey and in third countries in the exploration and development of hydrocarbons, and also approved the obligatory basis for contractual cooperation between the parties by supporting the supply of oil and oil products from the Russian Federation to the Republic of Turkey in the amounts established by commercial contracts with Russian oil companies.

Undoubtedly, the adopted international agreements, acting as one of the main forms of implementing measures to provide the states with their obligations, will contribute to strengthening interstate integration. The agreements are designed not only to increase the economic potential of the Russian Federation as an energy power and to avoid conflict situations between our state and foreign energy partners, but also to create a detailed regulated base of foreign trade contractual relations of a private nature. It can be confidently asserted that the process that was started helped to eliminate many of the difficulties that arose a few years ago when, due to the lack of clear mechanisms for regulating interstate relations in the delivery of energy products, in which, first of all, the participants of private legal relations suffered—the subjects of the national liability law.

Another problem of legislative nature attracts attention in the sphere of national ownership of land.

In accordance with the Federal Law "About State Registration of Rights to Immovable Property and Transactions with It" (Federal Law 1997), state registration of rights is possible only for land plots and real estate objects that are part of the enterprise as a property complex. According to the Law, the land plot is registered separately from the subsoil plot. According to the Law, the land plot is registered separately from the subsoil site.

Pevzner M.E., for example, insists on the necessary delineation in the law on the subsoil of the right of ownership of land from the right to use subsoil (Pevzner 2001).

10   Conclusion

In the Russian Federation, ownership of a land plot does not entail ownership of subsoil resources under it. Ownership of a land plot in Russia may be private, but ownership of subsoil is only state property. Therefore, experts point out that it is necessary to distinguish between ownership of land and the right to use subsoil. In other words, the owner of a land plot may not necessarily be a subsoil user. For this, there is a general procedure for granting subsoil for use, regardless of the ownership of land. But, for example, in the United States of America, as in some European countries, the owner of the land is identical to the owner of the bowels (Ageev 2008).

Despite this state of affairs in modern legal regulation, which makes one think about the need to weaken state control over the use of natural resources, we should recall the reforms of Catherine II in 1782 (the years of the reign of 1762–1796), pursuing a similar goal and successfully attaining it. In the laws of Catherine, previously repeatedly commented on in the studies of jurists, new rules for the use of natural resources were introduced. Thus, it was established that "the owner of the land is given the full right to all works, on the surface and in the waters of the earth, found, and all the metals, minerals and other minerals that are hidden in the bowels of the earth." Ownership of natural resources became inseparably linked with the right of ownership of land. Such a regime of legal regulation of nature management was maintained in Russia until 1917. The right of ownership of everyone in his estate extended to forests and all growths, water and the interior of the earth (Kalinin 2012).

Today in the Russian Federation in order to obtain a permit to develop new oil fields it is necessary to overcome a number of bureaucratic obstacles connected with the need for approvals with the owner of the subsoil—the state. For comparison, in the USA, which successfully implements shale oil production through innovative technologies, such activity is regulated privately. Legislative consolidation of a similar order in the Russian Federation would help accelerate the development of shale hydrocarbons and reduce the financial capital intensity of projects.

Thus, it should be concluded that in addition to the problems of economic and legal regulation of the development of oil shale deposits, there are a number of environmental risks, the onset of which, like the efficiency of their extraction, depends on technological problems. As already mentioned, the development of shale oil requires the use of innovative technologies in addition to modern drilling equipment and a large amount of electricity. Due to the fact that the extraction of shale oil is a fairly young and unconventional way among the extraction of natural resources, until now all the consequences of large-scale extraction of shale oil for the environment are not revealed. Ecologists warn of the grave danger of rock drilling, which can lead to contamination and, consequently, contamination of underground drinking sources, and also increases the threat of earthquakes in seismic regions.

Ecological threat is represented not only by the process of extraction of shale oil, but also by the method of conservation of the well. The current method does not guarantee that the residual oil can be isolated from the environment. As a result, contamination of fertile soils can occur.

In the science of ecological law, when researching the problems of rational use of natural resources, emphasis is often placed on the obligation to ensure the reproduction of natural resources. In this regard, despite the possible economic benefits, it is very important to maintain a balance of economic and environmental interests when developing new technologies for oil production, as well as other minerals.

Taking into mind the strategic importance for the Russian Federation of effective legal regulation of oil production through innovative technologies, it is obvious the need for legislative development of a number of legal, tax and environmental problems. In addition, there is a need for the formation of such a state energy policy in the field of subsoil use and management of the state subsoil fund, which will ensure sustainable, maximum effective and environmentally safe use of oil resources.

References

Ageev, R. V. (2008). Provision of land for geological study of subsurface resources. *Journal of Russian Law, 12,* 121–128.

Agreement. (2005). *On the organization, management, functioning and development of the common oil and gas market, (approved by the decree of the Government of the Russian Federation of September 26, 2005 no. 1499-p).* (Russian).

CNN Money. (2018). America could become oil king of the world in 2018, http://money.cnn.com/2018/01/03/investing/oil-us-russia-saudi-arabia-shale/index.html.

Dovbnya, V. B. (2012). Delimitation of powers between federal bodies and subjects of the Russian Federation in the field of possession, use and disposal of natural oil and gas resources. *Constitutional and Municipal Law, 8,* 48–50.

Federal Law. (1997). *On state registration of rights to real estate and transactions with it, (Adopted by the State Duma on July 21, 1997, no. 122-FL).* (Russian).

Federal Law. (2011). *On Ratification of the Agreement on the Organization, Management, Functioning and Development of Common Oil and Oil Products Markets of the Republic of Belarus, the Republic of Kazakhstan and the Russian Federation, (Adopted by the State Duma on July 1, 2011 no. 189-FL).* (Russian).

Gazprom Neft. (2017). *The project of Gazprom Neft for the study of the Bazhenov suite was given the status of a national one,* http://www.gazprom-neft.ru/press-center/news/1119467/.

GOST R 52104-2003. (2003). *National standard of the Russian Federation. Resource-saving (approved by the Decree of the State Standard of the Russian Federation on July 3, 2003 no. 235-st).* (Russian).

Grinets, I. (2014). The role of innovative development in unconventional hydrocarbon exploitation in the context of the shale gas revolution in the USA. *Journal of Social, Political, and Economic Studies, 39*(4), 436–466.

Inshakova, A. O. (2014). Innovative technologies of oil production: Tasks of legal regulation of cross-border management. *Bulletin of Volgograd State University. Series 5. Jurisprudence, 4* (25), 28–36.

Kalinin, I. B. (2012). Evolution of the legal regulation of the use of natural resources in Russia. *Russian Justice, 3,* 69–72.

Motin, V. V., & Trofimov, O. E. (2013). Problems of ensuring safety on water transport. *Administrative and municipal law, 6,* 607–611.

Muslimov, R. Kh. (2009). *Features of exploration and development of oil fields in a market economy: Proceeding allowance.* Fen Publishing House of the Academy of Sciences of the Republic of Tatarstan, Kazan.

Neftegaz. (2017). *The first in Russia. Gazprom Neft will apply a full cycle of technologies for the development of shale oil for the development of the Bazhenov suite,* http://neftegaz.ru/news/view/152699.

Noguera, J. (2017). The seven sisters versus OPEC: Solving the mystery of the petroleum market structure. *Energy Economics, 64,* 298–305.

Official website of the Ministry of Energy of the Russian Federation (2017). http://minenergo.gov.ru.

Perchik, A. I. (2006). Theoretical aspects of the formation of raw materials safety as a basic component of energy security. *Energy Law, 2,* 68–73.

Petrova, O. V. (2010). *Geological dictionary. In three volumes. T. 1. Izd. third, revised. and additional.* Publishing house VSEGEI, St. Petersburg.

Pevzner, M. E. (2001). Modern legal problems of subsoil use. *State and Law, 4,* 36–42.

Popondopulo, V. F., Gorodov, O. A., & Petrov, D. A. (2011). Renewable sources of energy in the electric power industry. *Energeticheskoe pravo, 1,* 23–29.

PwC: Audit and Assurance, Consulting and Tax Services. (2017). https://www.pwc.com/.

Salieva, R. N. (2013). Legal and environmental aspects in the use of primary energy sources within the framework of the Energy Strategy of Russia. *Jurist, 21,* 27–31.

Shakhverdiev, A. Kh., Dutov, D. V., & Shakhverdiev, E. A. (2009). Problems of harmonization of normative-legal aspects of the legislation on the subsoil use in Russia. *Neftyanoe khozyaystvo —Oil Industry, 10*, 20–23.

SNBC. (2018). US shale oil output poised to surge above 6.5 million barrels a day in February, https://www.cnbc.com/2018/01/16/us-shale-oil-drilling-poised-to-surge-in-february.html.

TASS. (2017). *Based on the materials of the Information Agency of Russia TASS*, http://tass.ru/ekonomika/4767246.

Tight oil expected to make up most of U.S. oil production increase through 2040, U.S. Energy Information Administration (2018). https://www.eia.gov/todayinenergy/detail.php?id=29932.

Tokarev, A. N. (2013). Taxation of the oil and gas sector of the Russian Federation: The role of regions. *International Accounting, 5*, 31–40.

U.S. Energy Information Administration (EIA). (2017). http://www.eia.gov/.

United Nations Convention. (1982). *On the Law of the Sea (UNCLOS) (Ratified by the Federal Law of the Russian Federation of February 26, 1997 no 30-FL)*.

Vedomosti. (2013). *Interview V. Yu. Alekperova*, http://www.lukoil.ru.

VYGON. Consulting. (2017). The main directions of the tax reform of the oil industry, http://vygon.consulting/products/issue-816.

Motivation of Energy Saving Within the Corporate Market Responsibility of Economic Entities

Elena G. Popkova, Oleg V. Inshakov and Aleksei V. Bogoviz

1 Annotation

Despite the unconditional importance of the role of the state as the guiding coordinator of the efforts of the society and the energy saving business, the key role in this process belongs to the economic entities that directly implement the energy saving strategy adopted at the national level. Therefore, it is important to involve them in the process and ensure that they are interested in its results.

In a modern market economy, involving little government intervention in socio-economic processes, external stimulus measures are usually characterized by low efficiency, both because of their artificial nature, and because of the complexity of control over their compliance. Therefore, the self-motivation of economic entities is a promising tool for implementing the national energy saving strategy.

Since market-making has already covered most of the spheres of the national economy in the countries of the world, including modern Russia, the most effective is the economic motivation that allows business entities to derive commercial benefits from energy conservation. In this chapter, the essence and specificity of

E. G. Popkova (✉)
Department of World Economics and Economic Theory,
Volgograd State Tecnical University, Volgograd, Russia
e-mail: erc@vstu.ru; 210471@mail.ru

O. V. Inshakov
Department of Economic Theory, World Regional Economics,
Institute of Economics and Finance, Volgograd State University, Volgograd, Russia

A. V. Bogoviz
Federal State Budgetary Scientific Institution "Federal Research Center
of Agrarian Economy and Social Development of Rural Areas – All Russian Research
Institute of Agricultural Economics", Moscow, Russia
e-mail: aleksei.bogoviz@gmail.com

© Springer International Publishing AG, part of Springer Nature 2019   95
O. V. Inshakov et al. (eds.), *Energy Sector: A Systemic Analysis of Economy,*
Foreign Trade and Legal Regulations, Lecture Notes in Networks and Systems 44,
https://doi.org/10.1007/978-3-319-90966-0_7

self-motivation of energy saving by business entities in the framework of their corporate market responsibility is studied.

2 Materials

Issues of motivation of energy conservation of business entities are reflected in publications (Park and Kwon 2017; Mori et al. 2015; Lv and Liu 2014; Zaripova et al. 2015; Steimer and Steimer 2010). The nature and characteristics of the market for corporate environmental responsibility of business entities identified in the work (Tan et al. 2017), (Bazarhanova et al. 2017). Institutional uncertainty of the ownership of power equipment to economic entities in modern Russia is being developed in scientific papers (Inshakova et al. 2017). Possibilities for concessional financing of energy saving projects in the form of syndicated loans are disclosed in publications (Inshakova et al. 2017).

3 Methods

Let us specify the categorical apparatus of research, and for we will analyze the essence of the process of motivation of energy saving within the corporate market responsibility of economic entities. On the basis of the content analysis of the scientific literature on this topic, we have identified two approaches to the interpretation of the essence of the process of motivation of energy saving within the corporate market responsibility of economic entities.

The first approach involves the consideration of self-motivation in focusing on the internal environment of the economic entity (enterprise). Under this approach uses a direct interpretation of the process of self-motivation, and it is assumed that the energy efficiency under its influence brought about by the internal motives and organizational structure to identify opportunities in energy conservation and implement them in the business practice of the enterprise, corporate culture, shaping and developing values of energy efficiency, etc.

Motives related to the influence of the external environment (globalization, market, state regulation, etc.) in the framework of this approach considered as imposed to the business entity. That is, the motivation of energy saving within the corporate market responsibility of economic entities involves the orientation on cultural values and does not involve the pursuit of commercial gain. The logic of this approach is explained in the works (Larrère 2017; Thörn and Svenberg 2017; Kim et al. 2017; Hirose et al. 2017; Wei et al. 2017; Testa and D'amato 2017; Kudlak and Kisiała 2017; Hook et al. 2017) etc.

In our opinion, this approach most accurately reflects the characteristics of self-motivation as a process initiated by the object of motivation (in the case of self-motivation the subject and the object of the motivation is the same) with the socio-psychological point of view but has limitations in practical application in the economy. This is because corporate social responsibility represents a specific phenomenon that emerged and continues to evolve under the strong influence of external economic factors.

4 Introduction

In the conditions of market economy, the main driving force of behavior of economic entities are their own or as they called by classics of economic thought—selfish—motives. In an effort to maximize the benefit (or use) economic agents can detect market signals and show a high flexibility, constantly adjusting their corporate strategy and tactics.

Under the influence of the trend of increased non-price competition and increasing consumer interest in the protection of environment and saving of natural resources, economic entities are increasingly implementing of energy saving practices in their economic activities. Since these initiatives are voluntary, they do not require the participation of the state that not only saves state budget, but also maintaining the balance of power in the market, without limiting the effect of natural and highly effective mechanism of competition.

This explains the priority of motivation of energy saving by economic entities, providing an increase of stability of development of socio-economic systems; maximize their efficiency due to the dominance of the market authorities. The purpose of this section is to study the expansion of self-motivation of energy saving within the corporate market responsibility of economic entities. To achieve it required the following tasks:

- Identification of key factors of motivation of energy saving within the corporate market responsibility of economic entities;
- Definition of the nature and logic of process of motivation of energy saving within the corporate market responsibility of economic entities;
- Development of practical recommendations for the development of self-motivation of energy saving within the corporate market responsibility of economic entities.

5 Types and Incentives of Activity of Economic Entities in the Field of Energy Conservation

For a long history of entrepreneurship, corporate responsibility has arisen only in the last century under the pressure of increased and aggravated the environmental problems, which resulted in stricter requirements of consumers and the state to the activities of enterprises. Therefore, from an economic point of view the second approach, considering the process of motivation of energy saving within the corporate market responsibility of economic entities in relation to the influence of external factors, more accurately reflects the economic substance of the process.

In the framework of this approach, the under-motivation of energy saving within the corporate market responsibility of business entities understood that the internal motives for energy saving under the influence of external factors. That is, to self-motivation refers to all the processes based on voluntary activity of a business entity in the field of energy saving. Corporate responsibility of business entities in this cases interpreted not as selfless initiative, and as a tool for commercial advantage.

Therefore, businesses tries to optimize their taxation, to attract and retain more customers, increase sales volume, to be able to enter new markets, etc. through the energy-saving process. Under this approach, motivation contrasted with extrinsic motivation of economic entities, involving the establishment of standards and requirements for energy efficiency.

For example, providing incentives for the activity in the field of energy conservation leads to self-motivation of economic entities and the tightening of legislation with the introduction of more stringent energy standards acts as an external motive for forcing business entities to implement energy savings. More details of this approach are considered in the research (Vicianová and Hronec 2017; Mulia et al. 2017; Gkorezis and Petridou 2017; Zhulavskyi et al. 2017; Karassin and Bar-Haim 2016; Xidonas et al. 2016; Miller 2016; Cai et al. 2016; Shubham et al. 2016), etc.

Based on our analysis of existing publications on the subject of the study revealed that although certain aspects of this problem are characterized by a high degree of scientific thoroughness, self-motivation energy saving businesses in relation to their corporate responsibility the market is not fully illuminated in the writings of contemporary scholars.

To identify factors of motivation of energy saving within the corporate market responsibility of economic entities, this study uses regression correlation analysis. Using these methods, we study the relationship between energy efficiency (y) and the level of ecological consciousness of the Russian population (x) through the calculation of the correlation coefficient and the determination of the sign of the estimated coefficient in the model, pair-wise linear regression.

Information-analytical base of the study was the sociological survey conducted by Levada center in 2015 about a representative nationwide sample of urban and rural population within 800 people 18 years old and older in 134 settlements of 46 regions of the country (distribution of answers given in percentage of the total number of respondents, together with data from previous surveys, the statistical error of these studies does not exceed 4.1%).

The authors also used materials of the report on the environmental responsibility of Russian companies in the rating of European countries by the cost of electricity, and the Government report on the state of energy conservation and energy efficiency in the Russian Federation in 2015. As indicators of efficiency in this work are used:

- The increase of the share of energy-efficient light sources in the lighting of economic entities, % (y1);
- The increase of the share of energy efficient buildings economic entities, % (y2);
- Reducing of the specific consumption of thermal energy in the ABs, % (y3);
- Reducing of the specific consumption of electrical energy on the BOF in ABs, % (y4);
- Reducing of the specific consumption of fuel and energy resources for manufacturing of products of economic entities, % (y5).

As potential factors influencing the motivation of energy saving within the corporate market responsibility of business entities, we have chosen the following:

- The growth of the perplexed (of interest) the environment the population of Russia, % (x1);
- Increase of demand for the products of environmentally responsible business (that is, the willingness of consumers to pay more for such a responsibility), % (x2);
- Increase of the level of corporate environmental responsibility of business entities (level of ecological competition), % (x3);
- Increase in electricity prices for business, % (x4).

The initial statistical data for the analysis grouped in Table 1. To ensure the representativeness of the data they given at five-year intervals, whereas time series analysis is used the whole fifteen-year sample.

Table 1. Dynamics of indicators of efficiency of economic entities and potential factors of their motivation to conserve energy

Indicators	The values of the indicators			
	2000	2005	2010	2015
y1	1,56	3,24	5,18	8,12
y2	1,35	2,80	4,48	7,03
y3	1,14	2,37	3,79	5,93
y4	1,18	2,45	3,92	6,14
y5	1,95	4,05	6,48	10,15
x1	1,05	2,18	3,49	5,47
x2	1,18	2,45	3,92	6,14
x3	1,67	3,47	5,55	8,69
x4	1,55	3,22	5,15	8,07

Source compiled by the authors on the basis of: (Levada center 2015; Wittenberg 2015; RIA Rating 2016; Ministry of energy of the Russian Federation 2015)

6 Results

On the basis of the statistical information given in Table 1 we obtained the following results of correlation analysis (Table 2).

As we can see on the Table 2, all potential factors, in fact, have a significant direct (as all the estimated coefficients in the models pair-wise linear regression have the " + " sign) effect on efficiency of economic entities in modern Russia. It is noteworthy that the most significant factor, correlation is the most severe in all of the indicators of energy efficiency and the value of the estimated coefficient in the regression model the most high, was the factor of x4 increase in electricity prices for business.

On the second place stands the factor x1, reflecting the growth of the perplexed (of interest) of the environment of the Russian population. Despite the similarities with the factors x2 and x3, which may exhibit random or one-time preferences, this factor reflects the level of environmental awareness of consumers and therefore indicates their willingness to abandon the potential benefits of a product if its manufacturer shows environmental irresponsibility when there are analogues.

In other words, consumers ready to abandon the product properties for the protection of the environment. The factors x2 and x3 represent the preferred products environmentally responsible business in comparison with existing analogues. That is, consumers are willing to pay more for environmental protection. A strong influence of x1 factor on the motivation of energy conservation of economic entities in comparison with the factors x2 and x3 due to the low level of solvency of the Russian population, which makes consumers more willing to abandon potential non-financial benefits than spending additional funds.

This suggests that the basis of self-motivation energy savings of economic entities is primarily guaranteed direct commercial benefit from saving resources, and only after that, they consider the potential marketing benefits of corporate market responsibility. Based on this, we have developed an algorithm of self-motivation of energy saving by economic entities, reflecting the nature and the logic of the process (Fig. 1).

Table 2. Correlation analysis results

From the factors	The values of correlation coefficients of indicators of energy saving				
	y1 (%)	y2 (%)	y3 (%)	y4 (%)	y5 (%)
x1	98,31	99,14%	99,52	98,75	99,49
x2	97,18	94,25%	96,87	94,36	95,64
x3	96,32	95,12%	93,82	98,26	97,15
x4	99,57	99,85%	99,64	99,72	99,35

Source compiled by the authors

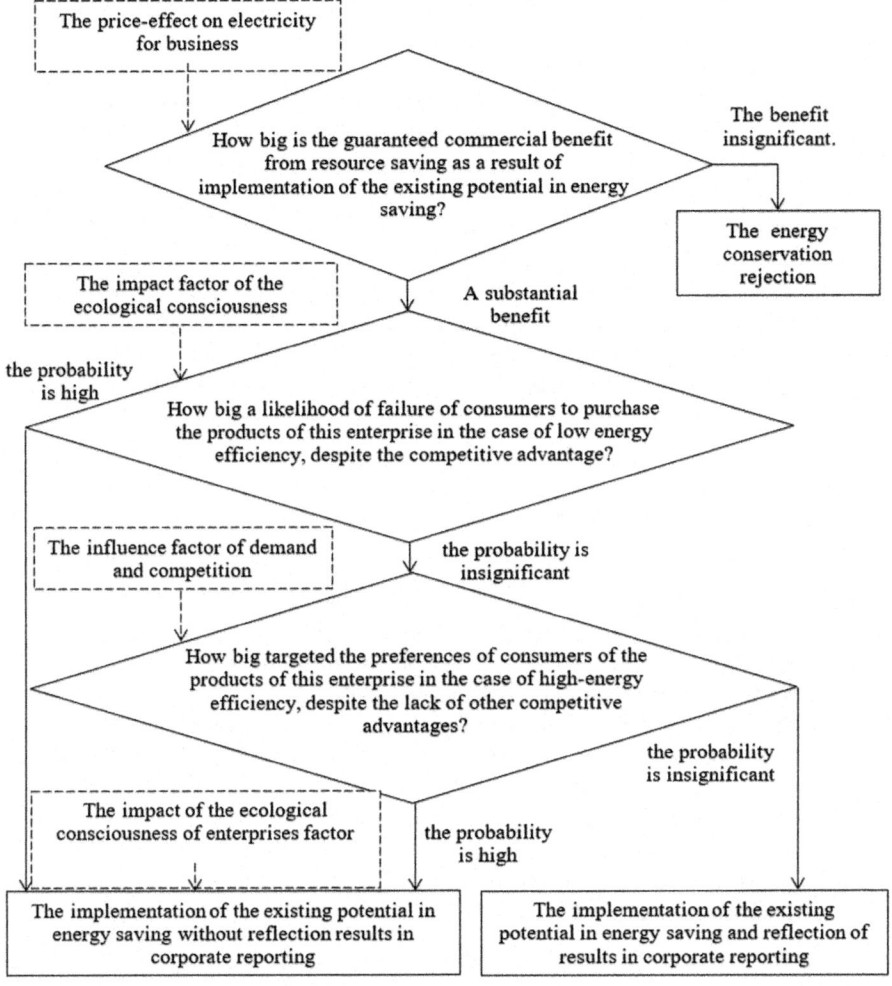

Fig. 1. The algorithm self-motivation power saving by business entities *Source* compiled by the authors

As can we see from Fig. 1, at first, farmers assess the value of guaranteed commercial benefit by saving resources as a result of implementation of the existing potential in energy saving. If such benefits are minor, they immediately refuse to energy efficiency, regardless of the influence of other factors. If the benefits are significant, farmers assess how great a likelihood of failure of consumers to purchase the products of this enterprise in the case of low energy efficiency, despite the competitive advantage.

If this probability is high, then there is the implementation of the existing potential in energy saving with reflection results in corporate reporting. Otherwise, evaluate how great a likelihood the preferences of consumers of the products of this enterprise in the case of high-energy efficiency, despite the lack of other competitive advantages. Here are the key factors of demand for products of energy-efficient companies and the level of market competition in the field of energy saving.

With little probability, there is a realization of the existing potential in energy saving without reflection results in corporate reporting, as it makes no sense, but it requires investment of time and resources. If the probability is high enough, it is the implementation of the existing potential in energy saving with reflection results in corporate reporting.

Only at the final stage of the process, self-motivation power saving by business entities influenced by the factor of the ecological consciousness of enterprises, that is non-profit motives associated with the desire of their leaders and employees to protect the environment through saving energy, and only in case if the results of this process are not reflected in corporate reporting. As otherwise, the commercial interests prevail over environmental initiatives.

Based on the identified entities and the logic of self-motivation energy saving we offer the following practical recommendations to facilitate the development of this process in the framework of the corporate market responsibility of business entities:

- Maintaining of prices of electricity for business on a level that they were high enough to ensure the economic feasibility of reducing them, but not so high as to contribute to the development of the shadow economy;
- Promotion of the development of innovations in the field of energy saving;
- The creation of favorable conditions for innovation in the field of energy saving;
- The development of ecological culture in the society: the formation of ecological consciousness in society and business, as well as the strengthening of the values of environmental protection and energy conservation in particular;
- Stimulating environmental competition of economic entities (their manifestation of corporate environmental responsibility with an emphasis on energy saving);
- Stimulating the demand on the products of environmentally responsible business (Inshakova et al. 2018).

The proposed recommendations focused on the state as a major subject management process-motivation of energy saving within the corporate market responsibility of economic entities. Despite the fact that the source of motivation are those entities in order to achieve this motivation and its increasing external intervention needed. In this regard, it is necessary to further improve the legal model of property relations in order to stabilize and develop the microeconomics of each business entity, including its self-motivation to save energy resources (Inshakova et al. 2017). In addition, a federal target program is needed, implemented by banks with state participation, under which particularly energy-intensive and environmentally-toxic business entities could receive preferential financing in the form of syndicated lending on a long-term basis (Inshakova et al. 2017).

7 Case Study

As an example from practice, we can introduce the major industrial enterprises of Russia. So, PJSC "Gazprom" pays the great attention to energy efficiency, setting targets for ourselves in the field of energy saving and completing them. According to the official statements of PJSC "Gazprom", in 2016 the savings of natural gas amounted to 12.8 billion m3, saving of electric energy—518.1 million kWh, saving heat energy—1259.8 thousand K.cal. and total energy savings amounted to 12 million tons ty. t. (Gazprom 2017).

The presence of a wide range of innovative opportunities to implement energy efficient production technologies and a significant proportion of energy costs determine significant guaranteed commercial benefit from saving resources as a result of implementation of the existing potential in energy saving. A special section of the "Environmental report" on the official website of the company and sub-mitting detailed plans and results of the activities in the field of environmental responsibility of business, highlighting energy efficiency in a separate paragraph reflects the high environmental consciousness of the consumers of its products.

The impact of such factors as demand for the products of environmentally responsible business (that is, the willingness of consumers to pay more for such liability) and the level of corporate environmental responsibility of business entities (level of ecological competition), is minimal because of PJSC "Gazprom" is a monopolist.

This suggests the absence or weak influence of the factor of environmental consciousness of the company. The dominance of commercial interests over environmental initiatives confirms the fact that the most expensive big economy of energy—thermal energy (25.4%) and electrical energy (20.6%), while savings the cheapest gas energy was the smallest and amounted to only 17.4% in 2016.

8 Conclusion

Thus, we can conclude that the motivation of energy saving, although it is one of the manifestations of corporate market responsibility of economic entities is more dictated by the market mechanism in the pursuit of their own commercial benefit. The environment and energy savings, in particular, this process may not worry about economic entities. Therefore, in the management of their initiatives in the field of energy saving, it is advisable to focus not on the illusory public benefit and on a clear private economic interest.

Formulated in this paragraph practical recommendations to facilitate the devel-opment of self-motivation of energy saving within the corporate market responsi-bility of business entities identify key areas of state management of the process, but does not reveal the essence of the instrumentation. In this regard, you need more focus on regulatory mechanisms of energy conservation in sustainable economic development.

Acknowledgements This study was funded by the RFBR according to the research project No. 18-010-00103 A.

References

Inshakova, A. O., Goncharov, A. I., & Sevostyanov, M. V. (2017). Institutional ambiguity of regulation of possessory relations in modern Russia. *Overcoming Uncertainty of Institutional Environment as a Tool of Global Crisis Management, 1*, 207–212.

Inshakova, A. O., Goncharov, A. I., Kazachenok, O. P., & Kochetkova, S. Y. (2017b). Syndicated lending: Intensification of transactions and development of legal regulation in modern Russia. *Journal of Advanced Research in Law and Economics, 3*(25), 838–842.

Inshakova, A. O., Frolov, D. P., Davydova, M. L., & Marushchak, I. V. (2018). Institutional factors of evolution and strategic development of general purpose technologies. *Espacios, 1*, 5.

Levada-center. (2015). Ecological situation. Retrieved June 21, 2017, from http://www.levada.ru/2015/06/18/ekologicheskaya-obstanovka/.

Ministry of Energy of the Russian Federation. (2015). State report on the state of energy conservation and energy efficiency in the Russian Federation in 2015. Retrieved June 21, 2017, from https://gisee.ru/upload/iblock/641/641965760ac81cfec023054748d34238.pdf.

Wittenberg, E. I. (2015). Environmental responsibility of Russian companies. Retrieved June 21, 2017, from http://dnevniki.ykt.ru/ivanshamaev/720400.

RIA Rating. (2016). The ranking of European countries according to the cost of electricity in 2016. Retrieved June 21, 2017, from http://www.riarating.ru/countries_rankings/20160701/630029918.html.

Bazarhanova, A., Kor, A.-L., & Pattinson, C. (2017). A belief rule-based environmental responsibility assessment for small and medium-sized enterprise. *FTC 2016—Proceedings of Future Technologies Conference*, 7821673, pp. 637–643.

Cai, L., Cui, J., & Jo, H. (2016). Corporate environmental responsibility and firm risk. *Journal of Business Ethics, 139*(3), 563–594.

Shubham, Charan, P., & Murty, L. S. (2016). Organizational adoption of sustainable manufacturing practices in India: integrating institutional theory and corporate environmental responsibility. *International Journal of Sustainable Development and World Ecology*, 1–12.

Gazprom. (2017). Environmental report of PJSC « Gazprom » for the year 2016. Retrieved June 21, 2017, from http://www.gazprom.ru/f/posts/36/607118/gazprom-ecology-report-2016-ru.pdf.

Gkorezis, P., & Petridou, E. (2017). Corporate Social Responsibility and pro-environmental behaviour: Organisational identification as a mediator. *European Journal of International Management, 11*(1), 1–18.

Hirose, K., Lee, S.-H., & Matsumura, T. (2017). Environmental corporate social responsibility: A note on the first-mover advantage under price competition. *Economics Bulletin, 37*(1), 214–221.

Hook, G. D., Lester, L., Ji, M., (…), Pope, C. G., & Van Der Does-Ishikawa, L. (2017). Environmental pollution and the media: Political discourses of risk and responsibility in Australia, China and Japan. Environmental Pollution and the Media: Political Discourses of Risk and Responsibility in Australia, China and Japan, c. 1–203.

Karassin, O., & Bar-Haim, A. (2016). Multilevel corporate environmental responsibility. *Journal of Environmental Management, 183*, 110–120.

Kim, H., Park, K., & Ryu, D. (2017). Corporate environmental responsibility: A legal origins perspective. *Journal of Business Ethics, 140*(3), 381–402.

Kudlak, R., & Kisiała, W. (2017). Environmental effects of social business responsibility—Institutional approach | [Środcmiskovve efekty społecznej odpowiedzialności biznesu—ujcie instytucjonalne: Wprowadzenie]. *Ekonomista*, 2017–January (3), pp. 243–263.

Larrère, C. (2017). *Environmental ethics: Respect and responsibility* (pp. 15–26). Rethinking Nature: Challenging Disciplinary Boundaries.

Lv, R. S., & Liu, H. R. (2014). Enterprises' participation in energy-saving activities' motivation mechanism model study. *Advanced Materials Research, 869–870*, 404–407.

Miller, L. F. (2016). Individual responsibility for environmental degradation: The moral and practical route to change. *Environmental Ethics, 38*(4), 403–420.

Mori, Y., Kobayashi, T., Anpo, Y., & Ohnuma, S. (2015). The long-term effects of intrinsic motivation on household energy-saving behavior: With actual energy use and self-reported behavior. *Research in Social Psychology, 31*(3), 160–171.

Mulia, P., Behura, A. K., & Kar, S. (2017). Corporate environmental responsibility for a sustainable future | [Odpowiedzialność środowiskowa biznesu dla zrównoważonej przyszłości]. *Problemy Ekorozwoju, 12*(2), 69–77.

Park, E., & Kwon, S. J. (2017). What motivations drive sustainable energy-saving behavior? An examination in South Korea. *Renewable and Sustainable Energy Reviews, 79*, 494–502.

Steimer, F. L., & Steimer, T. (2010). Green E-Business: Sustainable advancements in energy saving and pollution prevention (IEEE-DEST 2010). 4th *IEEE International Conference on Digital Ecosystems and Technologies—Conference Proceedings of IEEE-DEST 2010*, DEST 2010, 05610639, pp. 252–255.

Tan, S.-H., Habibullah, M. S., & Tan, S.-K. (2017). Corporate governance and environmental responsibility. *Annals of Tourism Research, 63*, 213–215.

Testa, M., & D'Amato, A. (2017). Corporate environmental responsibility and financial performance: Does bidirectional causality work? Empirical evidence from the manufacturing industry. *Social Responsibility Journal, 13*(2), 221–234.

Thörn, H., & Svenberg, S. (2017). The Swedish environmental movement: Politics of responsibility between climate justice and local transition. *Climate Action in a Globalizing World: Comparative Perspectives on Environmental Movements in the Global North*, pp. 193–216.

Vicianová, J. H., & Hronec, S. (2017). Environmental impact of application on the concept of corporate social responsibility in selected EU countries | [Środowiskowe konsekwencje wprowadzania koncepcji odpowiedzialności społecznej biznesu w wybranych krajach UE]. *Problemy Ekorozwoju, 12*(2), 79–88.

Wei, Z., Shen, H., Zhou, K. Z., & Li, J. J. (2017). How does environmental corporate social responsibility matter in a dysfunctional institutional environment? Evidence from China. *Journal of Business Ethics, 140*(2), 209–223.

Xidonas, P., Doukas, H., Mavrotas, G., & Pechak, O. (2016). Environmental corporate responsibility for investments evaluation: an alternative multi-objective programming model. *Annals of Operations Research, 247*(2), 395–413.

Zaripova, V. M., Petrova, I. Y., Kravets, A. G., & Evdoshenko, O. (2015). Knowledge bases of physical effects and phenomena for method of energy-informational models by means of ontologies. *Communications in Computer and Information Science, 535*, 224–237.

Zhulavskyi, A Yu., Smolennikov, D. O., & Kostyuchenko, N. M. (2017). Social and environmental responsibility strategies of business. *Naukovyi Visnyk Natsionalnoho Hirnychoho Universytetu, 3*, 134–139.

Regulatory Mechanisms of Energy Conservation in Sustainable Economic Development

Elena G. Popkova, Oleg V. Inshakov and Aleksei V. Bogoviz

1 Annotation

Sustainability is one of the most important principles for the development of modern economy. The importance and urgency of this is emphasized in the numerous works of domestic and foreign authors which serves as a prerequisite for the proclamation of a course for sustainable economic development in the countries of the world. At the same time the applied aspects of this issue remain poorly understood. Since theoretical models are not supported by practical recommendations concrete measures to ensure the sustainable development of the economy virtually taken.

Energy saving being a way to optimize the use of natural resources, allows balancing the interests of socio-economic development and preservation of the environment, which is the main idea of the concept of sustainability of the economic system. This chapter is devoted to the study of successful experience, features and prospects for energy saving in Russia and abroad, through the application of regulatory mechanisms of economic motivation of economic entities in the interests of sustainable economic development, as well as the development of practical solutions to this scientific problem.

E. G. Popkova (✉)
Department of World Economics and Economic Theory,
Volgograd State Technical University, Volgograd, Russia
e-mail: erc@vstu.ru; 210471@mail.ru

O. V. Inshakov
Department of Economic Theory, World and Regional Economics,
Science of the Russian Federation, Volgograd State University, Volgograd, Russia

A. V. Bogoviz
Federal State Budgetary Scientific Institution "Federal Research Center
of Agrarian Economy and Social Development of Rural Areas – All Russian Research
Institute of Agricultural Economics", Moscow, Russia
e-mail: aleksei.bogoviz@gmail.com

© Springer International Publishing AG, part of Springer Nature 2019   107
O. V. Inshakov et al. (eds.), *Energy Sector: A Systemic Analysis of Economy,*
Foreign Trade and Legal Regulations, Lecture Notes in Networks and Systems 44,
https://doi.org/10.1007/978-3-319-90966-0_8

2   Materials

Conceptual and practical questions of regulation of motivation for energy saving describes in detail in the studies of modern scholars such as: (Ma and Yu 2017; Hafezalkotob 2017; Li et al. 2016; Zhu and Ruth 2015; Xie 2015; Okubo 2013; Hu 2014; Park and Kwon 2017; Udalov et al. 2017; Mori et al. 2015; Lv et al. 2014; Fukuda and Munakata 2014; Joachain and Klopfert 2014) ect. Directions for improving the legal regulation of the corporate structure of economic entities are disclosed in scientific papers (Inshakova et al 2017). The issues of legal support for large-scale syndicated loan projects for financing long-term energy conservation programs are explored in publications (Inshakova et al. 2017).

3   Methods

To analyze the connection of energy conservation and sustainable development of economy in modern Russia, regression and correlation analysis method used. Using these methods the authors prepared a model pairwise linear regression of the formula: $y = A + Bx + C$, where A is a constant, B is the estimated coefficient; C is the standard deviation of error.

The author provides the economic interpretation of the essence of statistical models based on the values of the estimated coefficient B (the value has its sign and magnitude). Also calculated the coefficients of determination (r2) for the regression models reflecting the degree of correlation selected for analysis of indicators.

Information and analytical base for the study are the materials of official statistical reports of such authoritative international organizations as the Sustainable Development Solutions Network and the Sustainable Society Foundation for 2006–2016. To analyze selected the following indicators (Table 1):

- Indicator of energy intensity (Energy Use);
- The energy-saving in the economy rate (Energy Savings);
- An indicator of the amount of greenhouse gases (Greenhouse Gases);
- Indicator of intensity of use of renewable energy sources (Renewable Energy);
- Integral indicator of climate and energy (Climate & Energy);
- Indicator of ecological well-being (Environmental wellbeing);
- General index of sustainability of economic system.

4   Introduction

Sustainability is one of the most important principles of modern economic development. The importance and relevance of this moment is emphasized in numerous works by Russian and foreign authors, which is a prerequisite for the proclamation

Table 1. Dynamics of indicators of energy conservation and sustainable development of economy in modern Russia

Years	x2	x3	x4	x5	x1, points from 1 to 10	y1, points from 1 to 150	y1, points from 0,00 to 5,00
	Indicators of climate & energy, points from 1 to 10						
	Energy use	Energy savings	Green-house gases	Renewable energy	Climate & energy integral indicators	Environmental wellbeing	Sustainability index
2006	1,01	3,40	1,00	1,00	1,36	140,00	0,98
2008	1,00	2,67	1,00	1,00	1,28	144,00	0,76
2010	1,00	3,17	1,00	1,00	1,33	146,00	0,87
2012	1,00	4,30	1,00	1,00	1,44	139,00	1,15
2014	1,00	3,22	1,00	1,00	1,34	147,00	0,86
2016	1,00	4,36	1,00	1,00	1,44	144,00	1,16

Source compiled by authors based on materials (Sustainable Development Solutions Network 2017; Sustainable Society Foundation 2017)

of the sustainable economic development in the countries of the world. However, applied aspects of this issue remain poorly understood. Since theoretical models not supported by practical recommendations, specific measures to ensure sustainable development of the economy is virtually nonexistent.

Energy conservation as a way of optimizing the use of natural resources, allows balancing the interests of socio-economic development and environmental conservation, which is the main idea of the concept of stability of an economic system. The purpose of this Chapter is to study the features and prospects of energy conservation in Russia, through the application of regulatory mechanisms of economic motivation of economic entities to ensure a sustainable economic development, and the development of practical solutions to this research problem. This goal achieved through the following scientific and practical problems:

- The analysis of energy conservation and sustainable development of economy in modern Russia;
- The definition of the features and problems of application of regulatory mechanisms motivating energy efficiency in the sustainable development of economy in modern Russia;
- The development of practical recommendation to improve the regulatory mechanisms of motivation of energy efficiency in the sustainable development of economy in modern Russia.

5 Economic and Legal Methods for Increasing Energy Efficiency: Program Acts and the National Economic System

As a result of the regression and correlation analysis, we obtained the following results (Table 2).

As can be seen in Table 2, in modern Russia the relationship between the indicators of climate and energy and ecological well-being as weak as the sustainability index of the economic system. The most significant link is shown with:

- an integral indicator of climate and energy (Y1 growth, when X changes on the 1: 1,18, y2 – 1,14, r2(1) = 0.85 and r2(2) = 0.91),
- an indicator of energy intensity of the economy (Y1 growth, when X changes on the 1: of 1.02, y2 – −3,33, r2(1) = 0,71, r2(2) = 0.23),
- the energy-saving in the economy rate (Y1 growth, when X changes on the 16: 1.07, y2 – 1,24, r2(1) = 0,74, r2(2) = 0,98).

The calculations showed that reducing energy intensity of the economy, reduction of greenhouse gases and increase the usage of renewable energy sources in modern Russia practically have not addressed, because the relevant indicators has the minimum scores. As the same time, the energy efficiency in the economy rate is on the average level in the rating system of the world that confirms the efforts in this direction by the Russian government and business entities.

In modern Russia on basis of regulation of energy saving processes laid down in the Federal law of November 23 2009 N 261-FL On energy saving and on increasing energy efficiency and on amendments to certain legislative acts of the Russian Federation" and by the State program of the Russian Federation "On energy Saving and energy efficiency improvement for the period till 2020".

Table 2. The results of the regression and correlation analysis

Dependence	The regression model (A + Bx + C)	The coefficient of determination (r^2)
y1(x1)	0,12 + 1, 18x + 0,242	0,85
y1(x2)	0,25 + 1, 1,02x + 0,485	0,71
y1(x3)	0,10 + 1, 1,07x + 0,374	0,74
y1(x4)	1,21 + 1, 0,05x + 3,57	0,32
y1(x5)	2,85 + 1, 0,03x + 4,95	0,37
y2(x1)	0,25 + 1, 14x + 0,234	0,91
y2(x2)	4,77-3, 33x + 2,81	0,23
y2(x3)	0,11 + 1, 24x + 0,176	0,98
y2(x4)	1,51-2, 2,22x + 0,22	0,29
y2(x5)	1,75-2, 34x + 2,16	0,25

Source calculated by the authors

The program of the Russian Federation "On energy Saving and energy efficiency improvement for the period till 2020" provides the significant public investment to promote energy efficiency in Russia. Within the period from 2011 to 2020 planned to allocate in energy-saving 70 billion rubles from the Federal budget, 625 billion rubles from the Russian Federation subjects funds (territorial, regional and local budgets) and 8837 billion rubles from extra-budgetary funds. That is, the total volume of public investment over 10 years will be $ 9532 billion.

This program contains 8 subprograms, focused on energy-saving and energy efficiency increasing in the electricity sector, heating supply and in public utilities, industry, agriculture, transport, state (municipal) establishments and in sphere of rendering of services, in housing and in the constituent entities of the Russian Federation.

According to the results of their implementation by 2020 will be planned to provide saving of 1124 million tons of conventional energy resources. In particular, 630 billion kW/h of electric energy, 1500 Gcal of heating energy, and 17 million tons of petroleum products. Under current 2017 average consumer prices in Russia cost of electric energy is of 5.50 rub/ kW/h, heating energy −1750 rub/Gcal and the price of Urals oil mark −400 $/t or 24000 rub/t. That is, the total energy savings over 10 years will be 4135.5 billion.

In addition, it is planned to achieve a total costs economy of energy resources by all consumers about 9255 billion rubles, the savings of the budgets of all levels on energy about 530 billion rubles. It is also planned to achieve savings on the subsidies for the purchase of energy about 260 billion rubles. Furthermore is planned to obtain additional fees for the tax on profit of organizations by decreasing energy consumption of the business about 346 billion rubles. The income growth of energy exports is planned the amount of 2700 billion rubles (Decree of the Government 2010).

That is, total savings of funds of all businesses in the country by increasing the economy's energy efficiency is expected about 13091 billion rubles. According to our estimates, the expected economic efficiency of this program is quite high. Comparing total benefits (4135.5 + 13091 = 17226.5 billion rubles) and costs (9532 billion roubles) received at 17226.5/9532 = 1.81. So, the benefits are almost 2 times higher than costs. From an environmental point of view, the effectiveness can be assessed as a low—program saves about 3% of conventional energy in 10 years.

For the implementation of this program by Federal law of November 23 2009 № 261-FL "On energy saving and on increasing energy efficiency and on amendments to certain legislative acts of the Russian Federation" provides the following main regulatory mechanisms of motivation of energy saving (Federal law 2009):

- Information support: providing for all interested persons (economic entities in the electric power industry, in heating and public utilities, agriculture, transport, state (municipal) establishments and sphere of rendering of services in housing and in the constituent entities of the Russian Federation) information about the necessity, plans and results of energy conservation in Russia;

- The development of requirements (recommendations) about energy efficiency: the Declaration of energy efficiency in the national economic system and promotion of energy saving;
- Compilation of plans of energy saving, monitoring and control of the progress and results of their execution: definition of a framework of strategic guidelines of the programs in the field of energy efficiency in the national economic system;
- Promotion of the development of energy management: promoting energy management, promoting the training of specialists in the field of energy management (education and training);
- Stimulation of technical re-equipment of the economy: the promotion of voluntary and compulsory inspection of economic entities (in the electricity, heating and public utilities, industry, agriculture, transport, state (municipal) establishments and sphere of rendering of services in housing and in the constituent entities of the Russian Federation) on the subject of energy efficiency.

According to analytical information about the state of the fuel and energy complex of Russia in 2016 of the Analytical center under the Government of the Russian Federation in the domestic economic system in recent years positive dynamics of GDP energy intensity has been observed, as evidenced by the data in Fig. 1.

As can be seen in Fig. 1, 2016, the energy intensity of Russia's GDP calculated at 0.212 t.e / thousand $, a decrease of 11,6% which is lower than the 2005. For comparison, in 2016, the energy intensity of China's GDP is 28% is lower and is 0,151 t.e/ thousand $, USA is on 41% lower and amounts about 0,089 t.e./ thousand $, European OECD countries −58% less and is 0,037 t./ t.e. (U. S. Energy Information Administration 2017).

In contrast with other economic systems with a similar level of socio-economic development, we can notice that the efficiency in modern Russia is low. Moreover, the reduction of energy intensity of Russia's GDP in recent years is largely due to the declining in GDP under the impact of the economic crisis than energy intensity reducing actions.

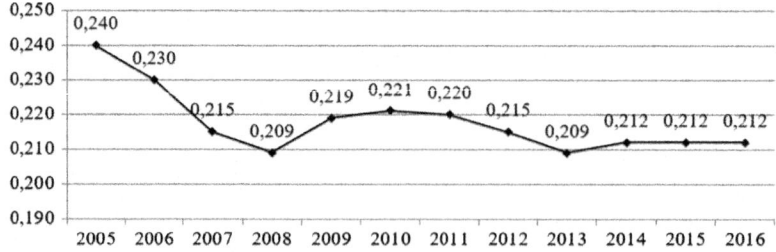

Fig. 1. Dynamics of energy intensity of Russia's GDP in 2005–2016, ton oil eq./thousand dollars
Source compiled by authors based on materials (Analytical center under the Government of the Russian Federation 2016)

Thus, a comprehensive analysis showed that the regulatory mechanisms of energy conservation used in modern Russia provide insignificant results in sustainable economic development. We identified the following key issues in the field of regulation of motivation of energy saving in Russia.

At first, the regulatory mechanisms of energy conservation used in modern Russia not aimed at the sustainable development of the economy. Based on the study of existing in the modern Russian legal documents we came to the conclusion that the primary purpose of state regulation of energy saving is the removal of environmental and financial benefits, expressed in a reduction of spending budgets and economic entities on the consumption and the increase of incomes of budgets of all levels, from improving energy efficiency. That is the most important landmark in the issues of energy saving in Russia is the problem of budget deficit, not the desire to achieve sustainable economic development.

Secondly, the regulatory mechanisms of energy conservation used in modern Russia involves the transfer of duties and responsibilities for the implementation of measures in the field of energy saving from the state to businesses. That is, the state itself is not engaged in active actions aimed at ensuring and/or promoting energy conservation.

Thirdly, special attention should paid to material incentives for the employees of economic entities themselves, since it is people who carry out actual actions, as result of which energy saving is implemented. Unfortunately, in the current period, employees considered owners and management of companies as a kind of asset—an integral part of the property complex. However, "referring people whose labor creates material and spiritual goods necessary to ensure the life of society, to the composition of the property complex, excludes the very possibility of manifesting in them motives for increasing labor productivity and raising the efficiency of social production. Recognition of them as hired workers, both private and public commercial entities and exclusion from production relations is the most harmful brake in the development of the productive forces of modern Russia" (Inshakova 2017).

One of potential measures is the development of energy standards. In 2006 was developed standards and labels for promoting energy efficiency in Russia. Despite the large-scale mission, it was a project of the Ministry of education and science. This is the evidence of its research focus, not practical and it was limited by the area of the city Moscow, and for the first 9 years of its implementation (the project ended in 2014) developed standards and recommendations on energy efficiency labelling is not widespread (Ministry of education and science of the Russian Federation 2006).

Overall control strategy of motivation of energy efficiency in the sustainable development of economy in modern Russia can be described as passive. This means that it provides the impact on the internal motives of economic entities.

In other words, in the framework of the current strategy is administered the promotion of implementation of voluntary initiatives of business entities in the field of energy saving.

It certainly determines its poor performance due to low interest of business entities in energy efficiency. This is evidenced by the results of sociological research conducted in 2016 in Russia, which showed that ecological problems are concerned only 13% of Russians. It is important to note that energy issues are far from first place in the structure of the environmental concerns of Russians (Levada center 2016).

The strategy of passive regulation of motivation of energy saving, implemented in modern Russia, is aimed at improving the literacy of economic agents in terms of energy savings. However, due to the low energy responsibility of economic entities is not conducive to the implementation of intensive measures and achieving meaningful outcomes. To achieve great results in the field of energy efficiency, we offer a transition to a strategy of active regulation of motivation of energy efficiency in the sustainable development of economy in modern Russia, which involves the creation of external motivations of businesses to the energy saving (Inshakova et al. 2018).

In this case, the initiator of the process of energy saving is the state that creates the conditions that compelled and motivated business entities to conserve energy. Under this strategy, we recommend to use the following regulatory mechanisms of energy conservation in sustainable economic development:

- The mechanism of standardization of energy consumption: the establishment of clear and obligatory high standards for power and control over their observance by all business entities with the introduction of a system of sanctions (fines—for the territories, and a liquidation—for companies);
- The mechanism of financial stimulation of energy efficiency: the provision of financial privileges to business entities, exhibiting high energy efficiency (subsidies, grants, tax benefits, etc.);
- The mechanism of marketing promotion of energy saving: the index of efficiency of economic entities, which allows those who are most active in the field of energy conservation to enhance their reputation in the market.

The proposed strategy presented graphically in Fig. 2.

As can be seen in Fig. 2, as a result of implementation of the developed strategy is achieved the high interest business entities in matters of energy efficiency, increase their energy literacy and responsibility. In particular, in the framework of granting financial privileges to large economic entities with high energy efficiency and participating in state programs of syndicated lending, it is advisable to annually reduce the interest rate on the syndicated loan received, covering this reduction by subsidies from the appropriations of the corresponding federal target program (Inshakova et al. 2017). This allows to create a modern culture of responsible energy consumption and subsequently to abandon of measures of economic motivation of energy conservation. That is in the terms of the proposed strategy we achieves not only a marked effect in the short term, but much greater effect in the long term, which is important advantage compared to the passive strategy of regulation and the basis of the strategy changing.

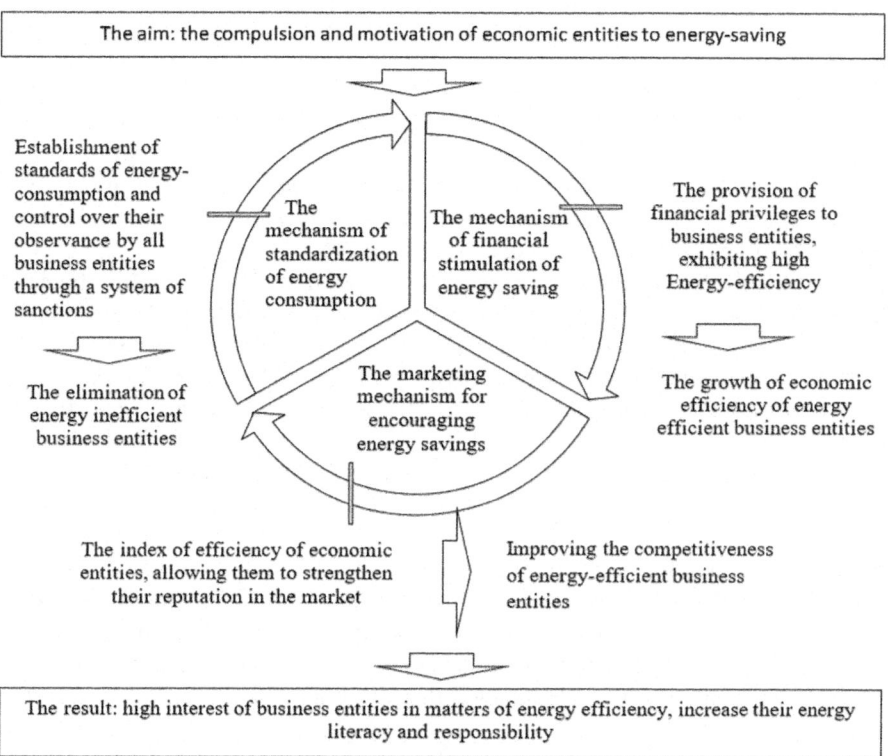

Fig. 2. The strategy of active regulation of motivation of energy efficiency in the sustainable development of economy in modern Russia *Source* compiled by the authors

6 Case Study

According to our forecast, as a result of implementation of the proposed strategy for active regulation of motivation of energy efficiency in interest of sustainable development of economy in modern Russia for 10 years much better results in energy conservation than the implementation of currently used strategies of passive management will be achieved. In particular, we expect that the savings of conventional energy resources representing 10–15%, reflecting a high ecological efficiency of the strategy.

On the basis of its implementation by 2030, it is planned to provide 5,000 million tons of conventional energy resources savings. In particular, the 3000 billion kW/h of electric energy, 7500 Gcal of heat energy and 85 million tons of petroleum products. At today's prices, total sum of energy savings over 10 years will be about 21000 billion/rub.

The savings of total cost of energy resources by all users estimated by us to be a 50000 billion/rub, the savings of the budgets of all levels for energy −1,500 billion/

rub. It is also planned to achieve savings on the subsidies for the purchase of energy in the amount of 15,000 billion/rub. It is planned to obtain additional fees for the tax on profit of organizations by decreasing energy consumption of the business in the amount of 1500 billion/rub. The growth of income from energy exports planned to be a 15,000 billion/rub.

The costs of standardization of energy consumption evaluated by us in 15000 billion/rub, the financial incentives for energy saving in 20000 billion/rub, on the marketing promotion of energy conservation in 5000 billion/rub.

According to our estimates, the expected economic efficiency of this strategy is higher than the efficiency of passive regulation strategy. So, the combined benefits amounts to 104000 billion/rub. (21000 + 50000 + 1500 + 15000 + 1500 + 15000), total cost 40000 billion/rub (15000+20000+5000). Comparing the benefits and costs we have 104000/40000 = 2, 6. That is, gain is 2.6 times higher than the costs.

As a result of implementation of the strategy is planned for 2030 to reduce energy intensity of Russia's GDP to 0.090 t. e./ thousand $, what is the level of European OECD countries. High economic efficiency of the developed strategy due to the fact that it involves the use of not only economic but also social mechanisms (including market mechanism) that allows to generate of entities responsibility approach to the consumption of energy. Combined with the economic motivation environmental motivation provides twice or even three times more results with a moderate growth of cost, which underlies the high efficiency of the presented strategy.

7 Conclusion

Thus, we can conclude that the relationship of energy conservation and sustainable development of economy in modern Russia quite clearly. Therefore, in the interest of increasing sustainability of the Russian economic system, it is advisable to focus on energy conservation.

Features of the application of regulatory mechanisms motivating energy efficiency in the sustainable development of economy in modern Russia, associated with the predominance of passive mechanisms of regulation (monitoring, information support, etc.) are problems in achieving high efficiency.

Proposed practical recommendations to improve the regulatory mechanisms of motivation of energy efficiency in the sustainable development of economy in modern Russia and a sound strategy of active regulation of motivation of energy efficiency in the sustainable development of economy in modern Russia can achieve a much greater economic and more importantly—environmental efficiency, ensuring long-term result.

Acknowledgements This study was funded by the RFBR according to the research project No. 18-010-00103 A.

References

Analytical center for the Government of the Russian Federation. (2016). FEC Russia—2016. Retrieved September 11, 2017, from http://ac.gov.ru/files/publication/a/13691.pdf.

Decree of the government of the Russian Federation from December of 27 December N 2446-p. To approve the enclosed state program of the Russian Federation « Energy Saving and energy efficiency for the period till 2020». (2010). Retrieved September 11, 2017, from https://rg.ru/2011/01/25/energosberejenie-site-dok.html.

Federal law of November 23, 2009 № 261-FL «On energy saving and on increasing energy efficiency and on amendments to certain legislative acts of the Russian Federation». (2009). Retrieved September 11, 2017, from http://base.garant.ru/12171109/3/.

Fukuda, M., & Munakata, J. (2014). *What kind of residents' motivations to improve lighting environment leads to energy-saving at home?* Indoor Air 2014—13th International Conference on Indoor Air Quality and Climate, c. 774–780.

Hafezalkotob, A. (2017). Competition, cooperation, and competition of green supply chains under regulations on energy saving levels. *Transportation Research Part E: Logistics and Transportation Review, 97,* 228–250.

Hu, D. J. (2014). Environment protection and energy saving in public private partnerships governance in the distributed energy generation. *Applied Mechanics and Materials, 508,* 236–242.

Inshakova, A. O., Frolov, D. P., Davydova, M. L., & Marushchak, I. V. (2018). Institutional factors of evolution and strategic development of general purpose technologies. *Espacios, 1,* 5.

Inshakova, A. O., Goncharov, A. I., Kazachenok, O. P., & Kochetkova, S. Y. (2017a). Syndicated lending: Intensification of transactions and development of legal regulation in modern Russia. *Journal of Advanced Research in Law and Economics, 3*(25), 838–842.

Inshakova, A. O., Goncharov, A. I., Mineev, O. A., & Sevostyanov, M. V. (2017). Amendments to the civil code of the Russian federation: Contradictions of theory and practice. *Contributions to Economics. Russia and the European Union: Development and Perspectives,* no. XIII, pp. 147–153.

Joachain, H., & Klopfert, F. (2014). Smarter than metering? Coupling smart meters and complementary currencies to reinforce the motivation of households for energy savings. *Ecological Economics, 105,* 89–96.

Levada Center. (2016). Poll: Most of all Russians are concerned about rising prices, poverty and unemployment. Retrieved September 11, 2017, from https://klops.ru/news/obschestvo/128377-opros-bolshe-vsego-rossiyan-bespokoyat-rost-tsen-bednost-i-bezrabotitsa.

Li, S., Liu, W., Li, Y., Wang, W., & Liu, F. (2016). Evaluation of energy-saving on peak load regulation scheme based on source-load coordination. Dianli Xitong Baohu yu Kongzhi. *Power System Protection and Control, 44*(12), 7–14.

Lv, R. S., & Liu, H. R. (2014). Enterprises' participation in energy-saving activities' motivation mechanism model study. *Advanced Materials Research, 869–870,* 404–407.

Ma, B., & Yu, Y. (2017). Industrial structure, energy-saving regulations and energy intensity: Evidence from Chinese cities. *Journal of Cleaner Production, 141,* 1539–1547.

Mori, Y., Kobayashi, T., Anpo, Y., & Ohnuma, S. (2015). The long-term effects of intrinsic motivation on household energy-saving behavior: With actual energy use and self-reported behavior. *Research in Social Psychology, 31*(3), 160–171.

Okubo, T. (2013). Energy-saving regulations and commodity prices. *Environmental Economics and Policy Studies, 15*(1), 93–132.

Park, E., & Kwon, S. J. (2017). What motivations drive sustainable energy-saving behavior?: An examination in South Korea. *Renewable and Sustainable Energy Reviews, 79,* 494–502.

Sustainable Development Solutions Network. (2017). The 2006-2016 SDG Index and Dashboards Report. Retrieved September 11, 2017, from http://sdsnyouth.org/sdg-index-2017/.

Sustainable Society Foundation. (2017). The 2006-2016 Sustainable Society Index. Retrieved September 2017, from http://www.ssfindex.com/.

The Ministry of education and science of the Russian Federation. (2006). Standards and labels for promoting energy efficiency in Russia. Retrieved September 11, 2017, from http://www.undp.ru/index.php?iso=RU&lid=2&cmd=programs&id=151.

U.S. Energy Information Administration. (2017). Global energy intensity continues to decline. Retrieved September 11, 2017, from https://www.eia.gov/todayinenergy/detail.php?id=27032.

Udalov, V., Perret, J., & Vasseur, V. (2017). Environmental motivations behind individuals' energy efficiency investments and daily energy-saving behaviour: evidence from Germany, the Netherlands and Belgium. *International Economics and Economic Policy, 14*(3), 481–499.

Xie, G. (2015). Modeling decision processes of a green supply chain with regulation on energy saving level. *Computers & Operations Research, 54,* 266–273.

Zhu, J., & Ruth, M. (2015). Relocation or reallocation: Impacts of differentiated energy saving regulation on manufacturing industries in China. *Ecological Economics, 110,* 119–133.

Energy Efficiency as a Driver of Global Competitiveness, the Priority of the State Economic Policy and the International Collaboration of the Russian Federation

Oleg V. Inshakov, Lyudmila Y. Bogachkova and Elena G. Popkova

1 Annotation

The global trend of reducing energy intensity and improving energy efficiency (EE) of the economy and reveals the connection of this process with the global competition of countries described below. Presents the status of state EE policy as a priority component of economic policy of the different governments and governing bodies of their unions. Given the comparative analysis of dynamics of indicators of economy of the Russian Federation and other countries. Convincingly the key role of technological energy saving in reducing energy consumption and improving EE in the economy. The necessity of the institutional, including legal, to ensure the participation of the Russian Federation in various forms of international collaboration is rejustified.

O. V. Inshakov
Department of Economic Theory, World and Regional Economics,
Science of the Russian Federation, Volgograd State University, Volgograd, Russia

L. Y. Bogachkova
Department of Applied Informatics and Mathematical Methods in Economics,
Institute of Management and Regional Economics, Volgograd State University,
Volgograd, Russia
e-mail: bogachkova@volsu.ru

E. G. Popkova (✉)
Department of World Economics and Economic Theory, Volgograd State Technical
University, Volgograd, Russia
e-mail: erc@vstu.ru; 210471@mail.ru

© Springer International Publishing AG, part of Springer Nature 2019
O. V. Inshakov et al. (eds.), *Energy Sector: A Systemic Analysis of Economy,
Foreign Trade and Legal Regulations*, Lecture Notes in Networks and Systems 44,
https://doi.org/10.1007/978-3-319-90966-0_9

2 Materials

The Analysis of the global transformation of energy and increasing the energy efficiency of the economy at the present stage of evolution of the global economic system is an extensive literature. It includes: Ayre (2013), Belogoryev et al. (2011), Bushuev et al. (2011), Bushuev et al. (2012), Shafranik (2015), IEA (2013a), IEA (2017), Makarov et al. (2016), EIA (2013b) etc.

A comprehensive theoretical and empirical studies of socio-economic effects of intensity of the economy reducing clearly proves in all spheres of human activity, combined with the institutional arrangement of property relations at the macro, micro, and nano levels of economic relations are the driver of economic development and the improvement of the well-being of the population. (Baatz 2015; Cambridge Econometrics 2015; EC Directorate-General for Energy 2016; IEA 2014a; IEA 2014b; IEA 2015; Vivid Economics 2013; Inshakova 2014; Inshakova et al. 2017 and others).

Extensive overview of contemporary literature on the study of the macroeconomic effects of EE growth are presented in the works: Naess-Schmidt, Hansen and von Below (2015), Saldanha et al. (2016).

Problems of development and state policy monitoring, which is implemented in the mainstream activities of governments of different countries, including Russia, as well as the governing bodies of their unions, highlighted in such publications as Braungardt et al. (2014), Improving Energy Efficiency Through Technology: Trends, Investment Behaviour and Policy Design (2011), EPA (2017a), EPA (2017b) and other. The issues of financing long-term projects within the framework of targeted government programs using syndicated bank lending are disclosed in scientific works: Inshakova et al. (2017) and others.

The need for international collaboration in the implementation of the plans and objectives of the EE policy emphasized in official documents and substantiated in scientific publications: G20 (2014), G20 (2016), Lesage et al. (2010), Mikkola et al. (2016) and other.

In Russia, as one of the countries with the highest energy intensity of the economy, EE increasing has the status of government economic policy priority.

So, it is considered as an important strategic objective, the need for a solution, which actualizes the problem of search, development and justification of tools and methods in managing the production and energy consumption.

In the present work in the context of comparative analysis of dynamics of energy intensity as one of the key performance indicators of the economic system of the Russian Federation and other countries, eloquently address the need for the relationship and EE's national competitiveness. It also makesthe case for necessity of expansion Russia's participation in international collaboration in EE sphere to preserve traditional and to gain new competitive advantages of the country. The author argues the relevance of institutional support for international cooperation of the Russian Federation and other countries in the energy sector and EE increase.

3 Methods

In this work used a general scientific methods: logical and comparative analysis, mathematical and statistical analysis, graphical modelling. The sources of statistical data are: the data of the RAS energy researches (The Energy Research Institute of the Russian Academy of Sciences (ERI RAS)), the World energy Agency (IEA), BP Statistical Review of World Energy, Statistical yearbook of global energy «Enerdata» (Global Energy Statistical Yearbook 2017).

4 Introduction

Increasing EE is one of the fundamental priorities and megatrends characterizing the development of the global economy in the XXI century. Under EE understands the optimal use of energy resources (fueland energy resources, FER) in achieved level of technological development and urgent environmental imperatives (Bushuev et al. 2012, pp. 20–26). The main EE indicator is specific consumption per unit of useful product in all spheres of human activity (economy, technology, and everyday life). In relation to the national economy, such indicator is EE's gross domestic product (GDP). The EE increase (or a decrease in energy capacity, or reduction of energy intensity) in the economic system of any level and scope can be, eider growth of volume of output at constant energy (constant energy consumption) or reduction in constant volume of the release; in the latter case we are talking about energy conservation as a private case of increasing EE.

5 The Global Trend of Energy Efficiency

Analysis of the main trends of the evolution of the global economic system presented in Bushuev et al. (2011), pp. 5–6, shows at present the global economy has neo-industrial process, comparable (comparable) in its significance and scale with the transition energy from the pre-industrial society based on biomass combustion, industrial energy based on fuels combustion. Energy in industrial phase, mainly based on the exploitation of mineral FER a large, multinational, vertically integrated and centrally managed companies. They were aimed at ensuring the greatest possible release of energy. Energy of neo-industrial phase is formed by the disintegration and decentralization of production processes, focused on the use of renewable energy sources, flexible automated control of energy systems with the help of modern information and computer technologies. The energy intensity of production processes is steadily declining, and increases EE economy.

Based on the above, the data given in the report of the President of the Russian Federation (Putin 2017), the reduction of energy intensity of national economies in modern conditions provided in the following ways:

- Substitution of exhaustible hydrocarbon resources, which include coal, oil, gas, renewable energy—like wind, solar, biowaste, geothermal (geothermal) sources, etc. On renewable energy sources now accounts for more than half of all world energy generating capacity input. By 2035, their share of world energy mix must increase from 15% to 23%, and in electricity generation (excluding hydropower) from the current 7–20%;
- Digitalization of energy complex. Fast processing of vast amounts of information and artificial intelligence, the introduction of «smart» power grids allows systematically analyze the production and consumption of energy in the long term significantly reduce the cost of energy resources, increase their efficiency and losses reduction (for example, «active house», which is supplied by Smart Grid electricity technology);
- The development of technologies of electric energy accumulation;
- Reorganization of the energy commodity markets, services and technologies;
- Formation of competitive substitutes for petroleum products used as motor fuel;
- Rapid development of nanotechnologies and their introduction in the energy sector to improve EE (Inshakova et al. 2017).
- Experts in a field of energy development note (Belogoryev et al. 2011, p. 14) that the innovative directions of transformation of energy become an independent sector with an annual turnover up to hundreds of billions of dollars and demonstrate high and sustainable growth in the countries-leaders of global development. In connection with the steady increase in EE by the middle of XXI century the transfer energy into a new qualitative state is expected.

IFA experts consider EE as a new «fuel» of the world economy (Ayre 2013). A retrospective analysis of the data for the years 1974–2010 is performed by them, in particular, showed that eleven OECD countries—Australia, Denmark, Finland, France, Germany, Italy, Japan, the Netherlands, Sweden and the UK, due to special measures aimed at increasing EE, managed to reduce energy consumption by 32 billion tons of oil equivalent (IEA 2013a). It secured the energy intensity decrease of their economies on average by 65% at the end of this period.

The declining trend of energy intensity (EI-energy intensity) in the global economy over the past quarter century is reflected in Fig. 1. In the period from 1990 to 2016, the worldwide energy in per unit of gross product consumption and dropped by 34% (from 0.219 koe/\$2005p to 0,144 koe/\$2005p). As shown by the linear trend (IE= $-0,002\ t + 0,220$), for the last 26 years, the energy intensity of the world economy by an average annual rate of 1% of the 1990 has been dropping.

The General trend of reduction in energy intensity evident in the BRICS countries, as well as in a number of countries with developed market economies (Fig. 2). Developing countries differ from developed large values of intensity, but exceed them in terms of improving EE.

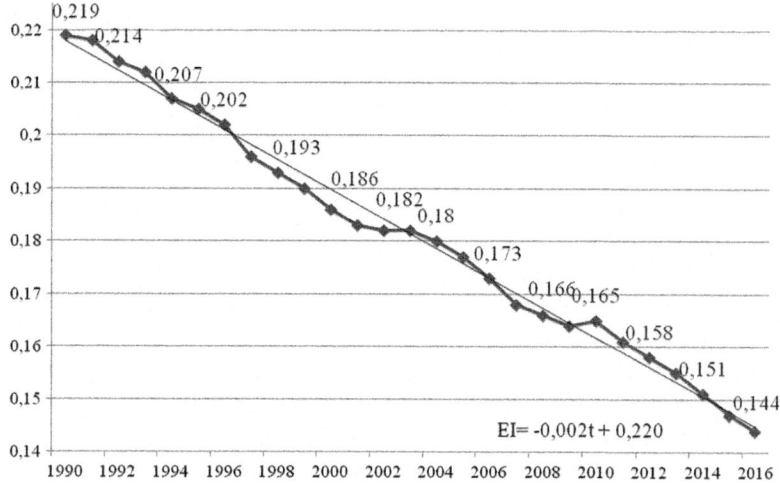

Fig. 1. World trend of energy intensity (EI) reduction, 1990–2016 (koe/$2005p—kg of oil equivalent per constant year 2005 USD at constant exchange rate) *Based on data* Global Energy Statistical Yearbook (2017). Enerdata

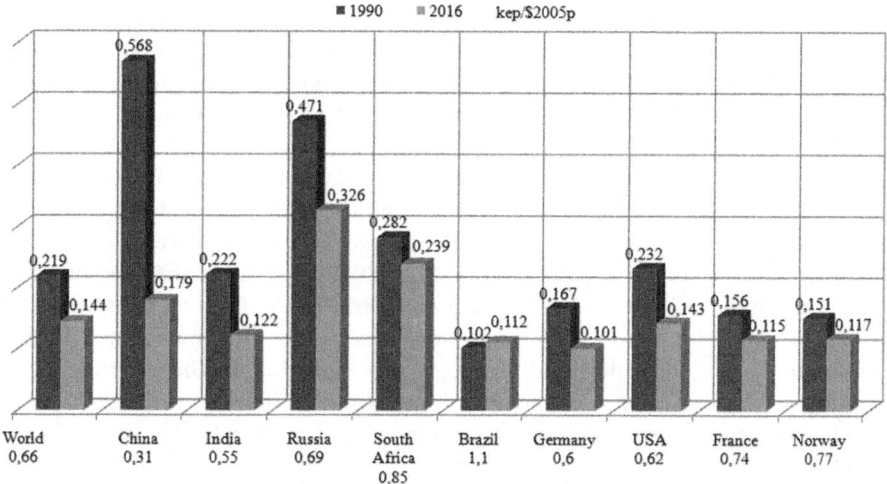

Fg. 2. The energy intensity of the economy in 1990 and 2016 in the BRICS countries and some countries with developed market economies (kep/$2005p—kg of oil equivalent/ US $ in 2005 prices in constant purchasing power parity) *Based on data* Global Energy Statistical Yearbook (2017). Enerdata. The bottom line—energy intensity index, showing which percentage of the base level of 1990 the energy consumption in 2016 was

In 1990 the average value of intensity for the BRICS countries is 86% higher than the corresponding average value for the USA, Germany, France and Norway (0,329 kep/$2005 in comparison with the 0,177 kep/$2005p). By 2016, the average value of intensity for the BRICS economies fell by 40%, and the corresponding average value for developed countries on 33% from the original values (0,196 kep/$2005 p in comparison with 0,119 kep/$2005p). Pace, or the reduction index of energy intensity for each country under consideration is shown in Fig. 2 in the bottom line (consumption value in 2016 toward to its value in 1990).

The highest rate of energy intensity pace (69%) demonstrates China: from 0,568 kep/$2005 p in 1990 to 0,179 kep/$2005 p in 2016. In the early period, the Chinese economy ranked the first place in terms of energy consumption in per unit of GDP. In 1990 the energy intensity of China was in 3.2 times higher than the average for developed countries; 1.2 times higher than in Russia; 5.57 times more than in Brazil—the least energy-intensive among the BRICS countries. By the end of this period Chinese economy has moved from first to third place among BRICS countries for energy consumption per unit of GDP, ahead of Russia and South Africa. In 2016, the energy consumption of China was only 1.5 times higher than the average for developed countries; 45% lower than in Russia; and only 1.59 times higher than in Brazil. Brazil only one of the BRICS countries energy intensity by 2016 has not decreased, but increased in comparison with the 1990 level (10%).

Figure 2 shows that the RF for the period showed the average rate of reduction in energy intensity (−31%), higher than in France (−26%), Norway (−23%) and South Africa (−15%) and better than in Brazil. However, despite this, in 2016 the Russian economy was the most energy intensive in the world (according to the Global Energy Statistical Yearbook 2017) that is associated with relatively high 1990 base level of this indicator. Currently, Russia energy saving and EE increasing is one of the most important strategic objectives (Ministry of Energy of the Russian Federation 2015).

New targets of further EE increase leading countries of the world set for them self too. Thus, the U.S. adopted a national action plan by 2025 halving the energy intensity of the added share of GDP in comparison with the volume and energy intensity of GDP in 2008 (EPA 2017a; EPA 2017b). Overall, the U.S. economy should become by 2020 on 25% less than in 2005. The EU has a Directive on EE, which provides for the energy reduction consumption by 20% relative to their consumption in 2007; the goal by 2020 full transition to the construction of buildings with zero energy consumption has been set. In Norway by 2025 in the private sector it is expected a full transition to electric vehicles (Braungardt et al. 2014).

EE raising, integral feature of which is the reduction of the energy intensity of the economy, ensuring competitiveness of national enterprises and their products, as well as the increase of the standard of living of the population, which is in particular, the reduction of energy costs in their production processes and final consumption.

For example, households in several countries significantly reduced their spending on energy consumption from 2000 to 2016 as a result of measures aimed at EE improving, as illustrated in Table 1.

Reduction of final consumption intensity processes is also clearly visible on the example of internal combustion engines. If even only 20 years ago car spent on

Table 1. Household annual energy expenditure savings due efficiency improvements since (2000–2016, USD per capita)

Countries\Indicators	The energy costs reuction in 2016 compared to 2000.	
	Absolute ($)	Relative (%)
Germany	580	30
France	470	30–35
United Kingdom	430	30
Japan	370	35
United States	140	10
China	60	25
Mexico	60	10

Based on IEA (2017)

Table 2. The long-term forecast of the relative energy intensity of some countries in the probabilistic scenarios of the global and national economies (relative to energy intensity of the world economy)

Country\Year	2020	2030	2040
Russia	1,63	1,67	1,62
China	1,27	1,22	1,12
India	0,91	0,89	0,87
USA	1,00	1,00	0,87
EU countries (in middle)	0,73	0,67	0,62

Based on the source Makarov et al. (2016)

100 km in an average of 12.2 L of gasoline, now at 8.5 L of gasoline, that is, the consumption decreased by 31% (Putin 2017).

The scientists of the Energy Research Institute jointly with the experts of the Analytical center under the Government of the Russian Federation is composed of the long-term forecast of the energy intensity of the Russian Federation and other countries (Makarov et al. 2016) (see Table 2).

As we can see in the Table 2, it is expected that in the future up to 2040 will remain relatively high energy intensity of the Russian economy. It will be exceed the world average in 2020–1.63%; in 2030–1.67; in 2040-1.62 times. By 2040 the expected energy of the Russian Federation will be higher than in China, India, USA and EU countries respectively: 1.45; 1,86; 1.86 and in 2.61 times. This development represents a serious threat to Russia, because with the EE's level of economy directly is directly related to the global competitiveness of countries (Government of the Russian Federation 2008; Shafranik 2016).

6 Energy Efficiency and Global Competitiveness

Insufficient EE and excessive energy consumption (excessive energyintensity) of individual production processes and the economy in General led to higher unit production costs and thus to underestimation of the competitiveness of end products of national producers in foreign markets. Therefore, global trends in energy efficiency reflected the desire of various countries to ensure the competitiveness and sustainability of national economy development.

National competitiveness (national competitiveness)—is a complex characteristic, reflecting on the one hand, the ability of the top companies in this country to compete successfully in international markets. On the other hand, the ability of the state to ensure the sustainability of economic development and welfare growth of the population due to the attraction of potentially effective resources and a high level of performance of the economy (Shvandar 2011). In the aspect of EE's definition of national competitiveness should emphasize the high level of economic efficiency imperative.

In the annual Report of the World Economic Forum reflects the global competitiveness index (World Economic Forum 2017), where the competitiveness of countries is defined as the toolbox of institutions, policies, and other factors that determine the level of productivity of the national economy (World Economic Forum 2016). Moreover, as Safiullin and Safiullin (2012) stress, the productivity of production factors is regarded as the main driver of national competitiveness. From this point of view, it deals with all the macroeconomic indicators of development of national economies, reflecting the increase or decrease in the productivity of factors of production. In this connection, the institutional certainty of property relations, including the long-term stability of the relationship between the ownership of energy facilities by economic entities (Inshakova et al. 2017) is of great importance for modern Russia.

It is generally believed, that the performance (productivity) as a measure of production efficiency, in relation with economy of the country can be treated in the same way as in relation to the industry or firm (Rayzberg et al. 1999; Academic Dictionary Online). Performance describes the output in per unit of resources level used and defined as the quotient of the volume of production by the amount of resources required to produce a given volume of production. In this, the resource costs considered as in aggregated form, and differentiated in physical units the volume of usage of individual factors.

Thus, the energy intensity of the economy, characterizing its level and EE defined as the ratio of GDP to the total costs of energy resources expressed in physical units, is one of the indicators of productivity of the economy, and reducing energy consumption and improving EE is priority of economic policy, contributing to increasing national competitiveness.

In modern conditions of competition in the markets for energy resources shifting to the area of advanced technologies of extraction and processing of raw materials (Shafranik 2015, p. 8). This trend is, according to M. Porter (Porter 1990), corresponds to the transition of most countries from the lower stage of competitiveness when a competitive advantage provided by the basic factors (excess minerals) to higher stages of competitiveness: investment-driven and innovation-driven. The main drivers of competitiveness is an active investing and attraction of investments; innovation. At the stage of innovations the lack of less high-tech resources is compensated by import and export is more high-tech. The country itself produces high technology and less dependent on their imports. This stage characterized by a high rate of EE increase.

Thus, the process of improving EE, reflecting the increase in performance the most important factor of production, emerging from fuel and energy resources (industrial valves) becomes the driver of growth in the competitiveness of national economies in the modern global economy. In connection with this EE policy becomes an integral and priority component of national economic policy, which is being developed and implemented in the mainstream activities of governments and governing bodies of their unions.

In Russia, the state EE policy currently defined by Energy strategy for the period up to 2030. But, Russian Federation has prepared the project of the new Energy strategy for the period up to 2035. In the new strategy, the reduction of energy intensity of Russian economy for the period from 2008 to 2014, stresses that the main contribution to this result made a recovery growth and structural shifts in the industry, and the influence of the most important technological factor offset by a decrease in the efficiency of the old worn-out equipment. In the future, a key role in reducing energy consumption and EE improving of the Russian economy is given a technological energy saving (energy-saving) (Government of the Russian Federation 2014; Ministry of Energy of the Russian Federation 2016).

7 The Energy Component of the Global Competitiveness of the Russian Federation

Russia, a country with a significant share of global hydrocarbon reserves, which is an exporter with revenues from energy trade—the source of a significant part of the state budget. In this regard, in the field of the Russian Federation energy for a long time was on the first, lower stage of competitiveness, while the most developed and successful developing countries have made great strides in the competitive advantages of a higher level gained.

To transition to the highest stage of global competitiveness, driven by investment and innovation in the Russian Federation pays great attention to R&d in energy sector, created favorable conditions for attracting private capital for their funding and for the development of state-private partnership in scientific-technical sphere. In Russia conditions for attracting large-scale investment in the development of new technologies are created, localization in the country of production of energy equipment and the added value of produced energy increase. At the present stage in Russia oil refineries modernized, a powerful oil and petrochemical complexes built: the Siberian and Tobolsk clusters, Eastern petrochemical company. Great work on the technological renovation of the energy companies is under way in the regions, especially in Tatarstan.

In the report of the President of the Russian Federation (Putin 2017) noted that in the framework of the state regulation of power industry of the country investment mechanisms support in generation based on renewable energy with a guaranteed return to the investor. As an example, the production of modern solar panels for domestic technologies and equipment for wind generation. Now, already, Russia has one of the most «clean» energy balances in a world in which nuclear and hydropower together account a 84%, as well as the production of electricity based on the gas and renewable energy resources usage. By 2035 it is planned to increase the share of «clean» electricity generation to 90%.

The development of the Russian Federation energy driven by investment and innovation, expressed in the economy restruction, reduction of losses in network infrastructure; energy-saving and digital technology implementation; specific fuel consumption in transportation and electricity generation reduction. This would reduce the energy intensity of the Russian economy and increase its global competitiveness.

8 International Cooperation in the Field of Energy Efficiency

Many countries—exporters of electricity and different fuels are not interested in international cooperation in EE improving, because coordinated efforts for energy saving in the importing countries can jeopardize the export potential of the energy suppliers and the volume of the sales markets reduction. Leaders of international cooperation are the countries with the most advanced developments and technologies, as the implementation of international standards in the EE field can enhance their market power and access for their corporations capacious markets. For example, 2/3 of the services of high-tech oil services in the world have only three companies from countries with developed market economies (Putin 2017).

For the past 40 years on the mega-level of the global economic system coordinating the activities of state support and promotion of EE actively implements the International Energy Agency (International Energy Agency, IEA). This independent global institution created in the mid-1970s is consistent with the OECD (organization for Economic Cooperation and Development, OECD), developing a common methodology, policies and criteria.

At the annual G8 summits (from 2014—G7) experts and heads of the IEA report on progress in this area and recommend a specific policy measures to improve EE of national economies (IEA 2010, p. 6). The consolidated complex of IEA recommendations for more than 25 spheres of economic activities in the 7 priority areas: cross-sectoral activity, buildings, appliances, lighting, transport, industry and power supply systems. In this, the Secretariat of the IEA emphasizes the need for a comprehensive implementation of all proposed measures, as no single measure will not help States in their efforts significantly improve the EE of national economies.

Today, most OECD members are countries with a developed market economy and high levels of energy efficiency. In the line with the growth of energy consumption in developing countries, the expected share of which in total world energy consumption in 2040 will be 65% (EIA 2013b), the problem of involvement of countries outside the OECD, in the international work to reduce the energy intensity of the economy mainstreamed.

To coordinate the activities of developed and developing countries in the field of EE in 2009 at the G8 Summit in L'Aquila (Italy), the International Partnership for Cooperation on energy Efficiency (IPEEC) was created, currently unifying 16 countries and the EU, which accounted for more than 75% of global GDP and energy consumption. Equal members of this organization are developed countries and the BRICS States, including Russia.

In 2014, the countries of the G-20 summit in Brisbane (Australia) adopted a Plan of action for EE promotion (G20 2014), the coordinator of the implementation of which IPEEC was assigned. The paper stresses that EE is a priority area for the G-20, as it can become a driver of economic growth, enhance energy security and exert a positive influence on the environment. Defined the scope of the joint work of G-20 countries in HER region, which include: transportation, appliances, buildings, industrial energy management, electric power plants, as well as policy measures to strengthen investment in EE.

In Russia at the EE macroeconomic level, on the one hand, has the status of a priority of the state economic policy aimed at providing solutions to common problems of innovative modernization of the economy. On the other hand, achieving a high EE can be considered as the imperative of neo-industrialization based on NBIC-convergence and the transition to a digital economy, which are set in the strategic documents long-term planning to 2020 and to 2030 (Government of the Russian Federation 2008; Ministry of Economic Development of the Russian Federation 2015). A quantitative indicator of the required level EE achievement of the Russian economy should be the reduction of the energy intensity of GDP by 40% between 2007 and 2020 (decree of the President of the Russian Federation of 4 June 2008 № 889; Government of the Russian Federation 2010).

To solve the EE problems requires a joint efforts and coordination among various countries. As emphasized the President of the Russian Federation (Putin 2017), «...The objective energy problems are that, the forward movement is possible only together, uniting our efforts.... Today it is important consistently remove the barriers to the free movement of energy resources, investment in their extraction and production, actively develop energy infrastructure, to work together to develop new technologies». To achieve new goals, Russia seeks to expand cooperation in different formats: BRICS, EEU, SCO, G20, OPEC, IPEEC, GECF (Gas Exporting Countries Forum) and also other structures and forums using to discuss issues of cooperation in EE improving.

Integration of regional markets, the emergence of new routes of energy supplies, including the Northern Sea Route and Silk Road, involves the expansion of participation of Russia in international collaboration in the field of energy and EE increase.

With the participation of foreign partners in the Russian Federation carried out a large-scale strategic investments in generation based on renewable energy sources, in high-tech projects have significance for the sustainable development of the entire Eurasian continent. For example, the production of liquefied natural gas, «Yamal LNG», gas pipelines «Nord stream 1» and «Nord stream—2»; «Turkish stream and Power of Siberia» construction. Capacity of the pipeline «Eastern Siberia—Pacific ocean» is increasing. Substantial part of the contracts receives financial support in the form of syndicated lending, in which the largest banks of Europe, China and Russia take part (Inshakova et al. 2017).

An important direction of international cooperation development in the sphere of EE is also a nuclear power. Russian company «Rosatom» is the world leader of nuclear power—plans to build 34 units abroad by applicants from different countries.

To meet the challenges related to the development of the global energy industry, regulation of global energy markets in a variety of countries, it is necessary to expand international cooperation, which requires adequate institutional support that will boost global competitiveness of many countries. Thus, the institutional, including legal, international collaboration is an important problem of practice and actual scientific task.

9   Conclusion

The current stage of evolution of the global economic system linked with a global trend of increasing EE of global and national economies. Over the past quarter of a century, global energy consumption per unit of gross product dropped by 34%. Despite the fact that the economy of the Russian Federation in the period 1990–2016, showed the average rate of reduction in energy intensity (−31%) in 2016 according to the IEA, it was the most energy intensive in the world. According to the ERI RAS forecast, the high level of energy intensity of Russian economy will continue in the future until 2040. High power consumption or low EE intensity poses a threat to Russia, as EE level is directly connected to its global competitiveness.

The main driver for global competitiveness of the national economy is the growth of its productivity, one measure of which is GDP energy intensity. Therefore, the reduction of consumption and improving EE economy considered as a priority and imperative components of the economic policy of most countries, including the Russian Federation.

In the modern energy markets, competition is shifted to the advanced technologies of extraction and recycling of raw materials, which corresponds to the transition of most countries to the competitiveness stages, driven by investment and innovation. Russia, competitive advantages of which for a long time ensured by hydrocarbons surplus stock, which corresponds to the lower stage of competitiveness, faces with a challenge of technological modernization of the economy and transition innovative development of the economy by 2030. Thus, conditions to reduce energy consumption, to improve energy efficiency and increase of global competitiveness in Russia will be created.

To solve the problems related to energy efficiency improving, Russia is seeking to expand international cooperation in different formats: BRICS, EEU, SCO, G20, OPEC, IPEEC, GECF (Gas Exporting Countries Forum), as well as other structures and forums for dialogue. Synergies and coordination of various countries in the global energy development and increasing their economies EE requires the formation of a coherent institutional framework with a common legal framework for international energy cooperation that will enhance their global competitiveness.

References

Academic Dictionary Online. Retrieved September 20, 2017, from https://dic.academic.ru/dic.nsf/econ_dict/11742. (Russian).

Ayre, J. (2013). *Energy efficiency is the world's most important "fuel," IEA Says In New Report*, Clean Technica, 25 October 2013. Retrieved September 10, 2017, from https://cleantechnica.com/2013/10/25/energy-efficiency-worlds-important-fuel-iea-says-new-report/.

Baatz, B. (2015). *Why everyone benefits from energy efficiency programs*, American council for an energy-efficient economy, 23 June 2015. Retrieved September 20, 2017, from http://aceee.org/blog/2015/06/why-everyone-benefits-energy.

Belogoryev, A. M., Bushuev, V. V., Gromov, A. I., Kurichev, N. K., Mastepanov, A. M., & Troitskiy, A. A. (2011). *Trends and scenarios of world energy development in the first half of the 21st century*, Energiya Publishing House, Moscow. Retrieved September 20, 2017, from http://www.energystrategy.ru/editions/trends.htm. (Russian).

Braungardt, S., Eichhammer, W., Elsland, R., Fleiter, T., Klobasa, M., Krail, M., Pfluger, B., Reuter, M., Schlomann, B., & Sensfuss, F. (2014). *Study evaluating the current energy efficiency policy framework in the EU and providing orientation on policy options for realising the cost-effective energy efficiency/saving potential until 2020 and beyond*, Report on behalf of DG ENER. Retrieved September 18, 2017, from Karlsruhe/Vienna/Rome, 19 September 2014, https://ec.europa.eu/energy/sites/ener/files/documents/2014_report_2020-2030_eu_policy_framework.pdf.

Bushuev, V. V., Belogoryev, A. M., Apolonskiy, O. Yu, Borgolova, E. A., & Timatkov, V. V. (2012). *Sustainable development of oil and gas companies: From theory to practice*, Energiya Publishing House, Moscow. Retrieved September 20, 2017, from https://istina.msu.ru/media/publications/book/afa/3a7/24577443/Ustojchivoe_razviie_neftegazovyih_kompanij_-_ot_teorii_k_praktike.pdf. (Russian).

Bushuev, V. V., Kurichev, N. K., Timatkov, V. V., & Troitskiy, A. A. (2011). *Russian electric power industry—2050 in the context of innovation development*, Institute of Energy Strategy, Moscow. Retrieved September 20, 2017, from http://www.energystrategy.ru/editions/en-21.htm. (Russian).

Cambridge Econometrics. (2015). *Assessing the employment and social impact of energy efficiency*, Final report. Volume 1: Main report, Cambridge Econometrics, Covent Garden, Cambridge, UK, November 2015. Retrieved September 4, 2017, from https://ec.europa.eu/energy/sites/ener/files/documents/CE_EE_Jobs_main%2018Nov2015.pdf.

EC Directorate-General for Energy. (2016). *The macroeconomic and other benefits of energy efficiency*, Final report, European Union, August 2016. Retrieved September 20, 2017, from https://ec.europa.eu/energy/sites/ener/files/documents/final_report_v4_final.pdf.

Enerdata: Intelligence + Consulting. *Global Energy Statistical Yearbook*. (2017). Retrieved September 25, 2017, from https://yearbook.enerdata.net/total-energy/world-energy-intensity-gdp-data.html.

Environmental Protection Agency. (2017a). *National action plan for energy efficiency*, United States Environmental Protection Agency. Retrieved September 20, 2017, from https://www.epa.gov/energy/national-action-plan-energy-efficiency.

Environmental Protection Agency. (2017b). *National action plan vision for 2025: A framework for change*, United States Environmental Protection Agency. Retrieved September 10, 2017, from https://www.epa.gov/energy/national-action-plan-vision-2025-framework-change.

Florax, R. J. G. M., de Groot, H. L. F., & Mulder, P. (2011). (eds.), *Improving energy efficiency through technology: trends, investment behaviour and policy design*, Edward Elgar Publishing, MPG Group Books, UK. Retrieved September 20, 2017, from https://econpapers.repec.org/bookchap/elgeebook/3830.htm.

G20 2014, *G20 Energy efficiency action plan: Voluntary collaboration on energy efficiency*, 16 November 2014, Australia. Retrieved September 4, 2017, from http://www.g20.utoronto.ca/2014/g20_energy_efficiency_action_plan.pdf.

G20 2016, *G20 Energy Efficiency leading programme (final version)*, China. Retrieved September 10, 2017, from https://ec.europa.eu/energy/sites/ener/files/documents/G20%20Energy%20Efficiency%20Leading%20Programme.pdf.

Government of the Russian Federation. (2010). The State Program of the Russian Federation "Energy saving and improving energy efficiency for the period up to 2020" (approved by the Decree of the Government of the Russian Federation of December 27, 2010 no. 2446-r). (Russian).

Government of the Russian Federation. (2014). *Agenda "Introduction of innovative technologies and modern materials in the fuel and energy sector" for the period up to 2018 (approved by the Decree of the Government of the Russian Federation on July 3, 2014 no. 1217-r)*, RG.RU. Retrieved September 20, 2017, from https://rg.ru/2014/07/08/tek-site-dok.html. (Russian).

Government of the Russian Federation. (2017). *The conception of long-term socio-economic development of the Russian Federation for the period up to 2020 (approved by the Decree of the Government of the Russian Federation of 17 November 2008 no. 1662-r)*, Information Legal Portal Garant. Retrieved September 10, 2017, from http://base.garant.ru/194365/. (Russian).

Inshakova, E. I. (2014). Development of alternative energy on the basis of nanotechnologies: Projected effects for the Russian economy. *Science Journal of Volgograd State University Global Economic System, 5*(28), 80–89. (Russian).

Inshakova, A. O., Frolov, D. P., & Davydova, M. L. (2017a). The institutional factors of strategic development and the tactical regulation of nanotechology. *European Research Studies Journal, 3,* 588–606.

Inshakova, A. O., Goncharov, A. I., & Sevostyanov, M. V. (2017). Institutional ambiguity of regulation of possessory relations in modern russia. *Overcoming Uncertainty of Institutional Environment as a Tool of Global Crisis Management, 1,* 207–212.

Inshakova, A. O., Goncharov, A. I., Kazachenok, O. P., & Kochetkova, S. Y. (2017c). Syndicated lending: Intensification of transactions and development of legal regulation in modern Russia. *Journal of Advanced Research in Law and Economics, 3*(25), 838–842.

International Energy Agency. (2010). *Progress in the implementation of energy efficiency policy in the G8 countries: Russia in the focus*. Retrieved September 20, 2017, from http://www.iea.org/Papers/2009/eer_ru.pdf. (Russian)

International Energy Agency. (2013a). *Energy efficiency: Market report*, IEA Publications, France, October 2013. Retrieved September 20, 2017, from http://www.iea.org/publications/freepublications/publication/EEMR2013_free.pdf.

International Energy Agency. (2013b). Future world energy demand driven by trends in developing countries, December, 2013, Energy Information Administration, USA. Retrieved September 10, 2017, from https://www.eia.gov/todayinenergy/detail.php?id=14011.

International Energy Agency. (2014a). *Capturing the multiple benefits of energy efficiency: Roundtable on industrial productivity and competitiveness: discussion paper*, IEA Headquarters Monday, 27 January 2014, Paris, France. Retrieved September 20, 2017, from https://www.iea.org/media/workshops/2014/eeu/industry/IEA_Industrialnonenergybenefitsbackgroundpaper_FINAL.pdf.

International Energy Agency. (2014b). *Energy efficiency: A key tool for boosting economic and social development*, IEA, Berlin, 9 September 2014. Retrieved September 10, 2017, from https://www.iea.org/newsroom/news/2014/september/energy-efficiency-a-key-tool-for-boosting-economic-and-social-development.html.

International Energy Agency. (2014c). *Capturing the multiple benefits of energy efficiency*, IEA Publications, 11 October 2015. Retrieved September 10, 2017, from https://www.iea.org/publications/freepublications/publication/Multiple_Benefits_of_Energy_Efficiency.pdf.

International Energy Agency. (2017). *Energy efficiency 2017*. Retrieved September 20, 2017, from https://www.iea.org/publications/freepublications/publication/Energy_Efficiency_2017.pdf.

Lesage, D., de Graaf, T. V., & Westphal, K. (2010). G8 + 5 collaboration on energy efficiency and IPEEC: Shortcut to a sustainable future?. *Energy Policy. 38*(11), November 2010, pp. 6419–6427. Retrieved September 20, 2017, from https://www.sciencedirect.com/science/article/pii/S0301421509007265.

Makarov, A. A., Grigoryev, L. M., & Mitrova, T. A. (eds.). (2016). *Forecast of the development of the world's and Russia's energy—2016,* Institute of Energy Research, RAS & Analytical Centre at the Government of the Russian Federation, Moscow. Retrieved September 1, 2017, from http://ac.gov.ru/files/publication/a/10585.pdf. (Russian).

Mikkola, M., Jussila, A., & Ryynänen, T. (2016). Collaboration in regional energy-efficiency development. *International conference on Smart and Sustainable Planning for Cities and Regions (Results of SSPCR 2015)*, Springer International Publishing, Switzerland, pp. 55–66.

Ministry of Economic Development of the Russian Federation. (2015). *Forecast of the long-term social and economic development of the Russian Federation for the period up to 2030 (developed by the Ministry of Economic Development of the Russian Federation),* Information Legal Portal Garant. Retrieved September 20, 2017, from http://base.garant.ru/70309010/. (Russian).

Ministry of Energy of the Russian Federation. (2015). *State report on energy saving and energy efficiency in the Russian Federation in 2014*. Ministry of Energy of the Russian Federation. Retrieved September 1, 2017, from https://minenergo.gov.ru/node/5197. (Russian).

Ministry of Energy of the Russian Federation. (2016). *Forecast of scientific and technological development of the fuel and energy complex of Russia for the period up to 2035 (approved by Minister of Energy of the Russian Federation A.V. Novak on October 14, 2016).* Retrieved September 1, 2017, from https://www.minenergo.gov.ru/node/6365. (Russian)

Naess-Schmidt, S., Hansen, M. B., & von Below, D. (2015). *Literature review on macroeconomic effects of energy efficiency improvement actions*, D6.1 report, COMBI, Copenhagen, September 2015. Retrieved September 20, 2017, from https://combi-project.eu/wp-content/uploads/2015/09/D6.1.pdf.

Porter, M. E. (1990). *The competitive advantage of the nations*. New York: MacMillan Press.

Putin, V. V. (2017). *Speech at the plenary session "Energy for Global Growth" of the First International Forum on Energy Efficiency and Energy Development "Russian Energy Week",* 4 October 2017. Retrieved September 10, 2017, from http://kremlin.ru/events/president/news/55767. (Russian).

Rayzberg, B. A., Lozovskiy, L Sh, & Starodubtseva, E. B. (1999). *Modern Economic Dictionary* (2nd ed.). Moscow: INFRA-M. (Russian).

Safiullin, M. R., & Safiullin, L. N. (2012). Competitiveness of Russia: The view of the World Economic Forum (review article). *Electronic economic bulletin of Tatarstan, (2–3)*, 27–41. Retrieved September 4, 2017, from https://elibrary.ru/download/elibrary_21384416_77564217.pdf. (Russian).

Saldanha, G. C., Gouvea da Costa, S. E., & de Lima, E. P. (2016). Energy efficiency frameworks: A literature overview. *27th Annual Conference Proceedings: Production and Operations, Management Society (POMS)*. Retrieved September 20, 2017, from https://www.pomsmeetings. org/ConfProceedings/065/Full%20Papers/Final%20Full%20Papers/065-0440.pdf.

Shafranik, Yu. K. (ed.) (2015). *Global energy and geopolitics (russia and the world)*, Energiya publishing house, Moscow. Retrieved September 1, 2017, from http://www.energystrategy.ru/editions/docs/global_energy.pdf. (Russian).

Shafranik, Yu K. (2016). Global energy changes and Russia. A new map of the world energy space. *Energy Policy, 3*, 3–12. (Russian).

Shvandar, K. V. (2011). Modern trends in the formation of international competitiveness of the national economy. Ph.D. thesis, Moscow State University, Moscow. (Russian).

Vivid Economics. (2013). *Energy efficiency and economic growth: Report prepared for The Climate Institute*, June 2013. Retrieved September 20, 2017, from http://www.climateinstitute.org.au/verve/_resources/Vivid_Economics_-_Energy_efficiency_and_economic_growth_June_2013.pdf.

World Economic Forum. (2016). *Methodology: The 12 pillars of competitiveness*, The Global Competitiveness Report (2014-2015). Retrieved September 20, 2017, from http://reports.weforum.org/global-competitiveness-report-2014-2015/methodology/.

World Economic Forum. (2017). *The Global Competitiveness Report 2017–2018*. Retrieved September 4, 2017, from https://www.weforum.org/reports/the-global-competitiveness-report-2017-2018. (Russian)

The Transformation of the Global Energy Markets and the Problem of Ensuring the Sustainability of Their Development

Oleg V. Inshakov, Lyudmila Y. Bogachkova and Elena G. Popkova

1 Annotation

Based on the analysis of aggregated indicators of market conditions of global markets characterized the modern state and the transformation of the market environment, in territorial aspect, short-term and medium-term perspective. The authors identified the main factors of transformation of the global energy markets, and outlined the prospects of their development, including issues of geopolitics and global energy security. In this article supported the urgency of the problem of sustainability of global energy markets, including problems of legal regulation of world trade energy products. Also this article presents the possible solutions to these problems based on international cooperation and reveals an important role of Russia as the largest supplier of energy resources to global energy markets.

O. V. Inshakov
Department of Economic Theory, World and Regional Economics,
Science of the Russian Federation, Volgograd State University, Volgograd, Russia

L. Y. Bogachkova
Department of Applied Informatics and Mathematical Methods in Economics,
Institute of Management and Regional Economics, Volgograd State University,
Volgograd, Russia
e-mail: bogachkova@volsu.ru

E. G. Popkova (✉)
Department of World Economics and Economic Theory,
Volgograd State Tecnical University, Volgograd, Russia
e-mail: erc@vstu.ru; 210471@mail.ru

2 Materials

Annual reviews devoted to the analysis of current data on the status and main trends of changes in world energy markets and forecasts of their further development published by international and national specialized organizations, companies, academic institutions and media. For example: the International Energy Agency (IEA), British Petroleum (BP), The Energy Research Institute of the Russian Academy of Sciences (ERI RAS), Enerdata, Bloomberg and many others.

Long-term tendency and prospects of transformation of the global energy markets in terms of the various long-term scenarios of development of world and Russian economy have consistently been the focus of numerous studies by Russian scientists (Belogoryev et al. 2011; Shafranik 2015; Bushuev et al. 2016; Makarov et al. 2016, etc.). Most major energy projects are realized within the framework of long-term, complex multilateral transactions that receive financial support in the form of syndicated lending, the peculiarities of which are discussed in publications (Inshakova et al. 2017).

In 2014 for the global energy markets a period of exceptional volatility and instability that threatens global energy security set in, and its widely discussed in the literature (WEC 2016; Ivanov and Matveev 2017; Makarov et al. 2016; Faucon 2016; Faucon et al. 2016, Husain et al. 2015; RIA Nakanune.RU 2016; Slav 2016, etc.)

The present work involves substantiation of necessity of clarification of a well-known concept of energy security (Sapir 2006) by including the sustainability imperative of global energy markets. The authors substantiates the important role of Russia as major energy power in the sustainable development of world and national energy markets. The implementation of this role will require broad cooperation with different countries in energy, therefore a big work on development of normative-legal base in the sphere of international law has been mainstreamed. At the same time, it is necessary to stabilize the legal regulation of the relations of economic entities at the domestic level, the relevant recommendations are published in the works (Inshakova 2017).

3 Methods

Methods of statistical analysis, generalization and abstraction, analysis and synthesis, graphic modeling were employed. Used sources of statistical data were: the data of the RAS energy researches (The Energy Research Institute of the Russian Academy of Sciences (ERI RAS)), the World energy Agency (IEA), BP Statistical Review of World Energy, Statistical yearbook of global energy « Enerdata » (Global Energy Statistical Yearbook 2017), electronic media for professional participants of the financial and energy markets (Bloomberg.com; Ereport.ru).

4 Introduction

The modern condition of the global energy industry is characterized by a change in technological structure and extreme aggravation of geopolitical contradictions between the countries-producers and consumers of energy resources. The development of world energy markets are characterised by a high degree of uncertainty and risk, dynamic changes in market conditions. They are expressed in the following: « the coup » in the value of world production and consumption of energy; the emergence of fundamentally new goods and services; the restructuring of the global energy balance; the reorganization of the global energy trade flows; significant changes in market positions of the countries-exporters and importers of energy resources.

5 Current State of Global Energy Markets According to the Data for 2016

Due to the significant differentiation of countries a natural supply of energy resources as well as volumes of their consumption, various states are divided into net exporters and net importers of energy products (coal, gas, oil, electricity and biomass). Such grouping is based on the indicator of the balance of foreign trade in energy, numerically equal to the difference between the import and export of primary energy. For a net exporter, this indicator is negative and for a net importer is positive. For geographical or geopolitical zone trade balance in energy is the sum of the respective balances economic areas included in this zone (Enerdata 2017). Global trade balance for 2016 is shown in Fig. 1.

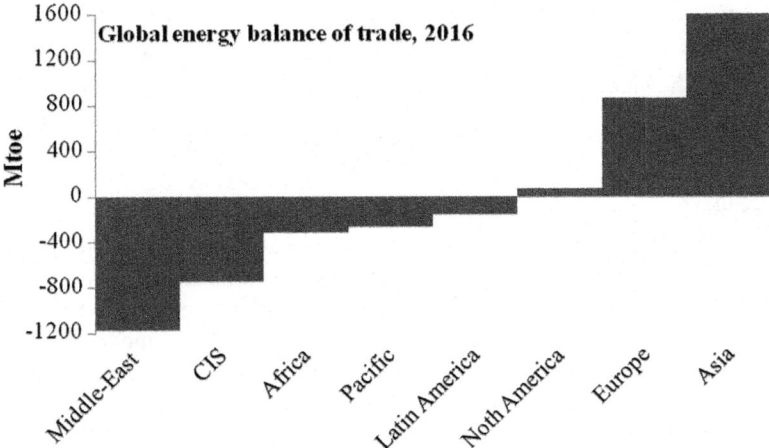

Fig. 1 Global Energy balance of trade, 2016. Includes coal, gas, oil, electricty, heat and biomass (Unit: Million tonnes oil equivalent (Mtoe). The trade balance is the difference between imports and exports. The balance of a net exporter appears as a negative value (−). The balance of geographic and geopolitical zones is simply the sum of the trade balance of all the countries

Russian energy balance of trade, 2016

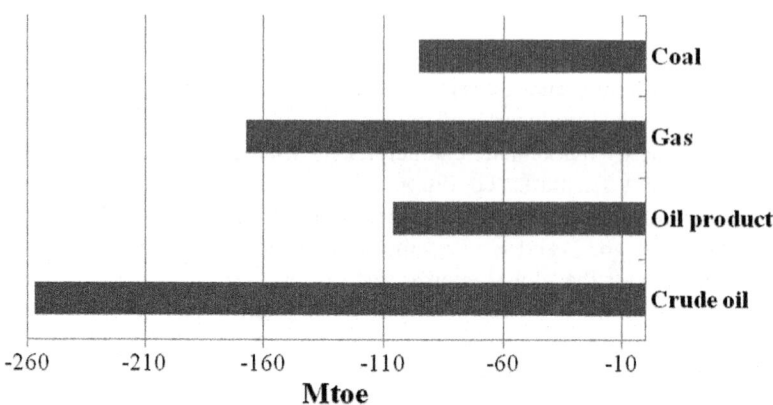

Fig. 2 Russian Energy balance of trade (the difference between imports and exports), 2016. (Unit: Million tonnes oil equivalent (Mtoe))

The main net exporters of energy resources are Russia, Saudi Arabia, Australia, Indonesia, Norway and Kuwait, the largest is Russia. Russian trade balance in energy illustrated in Fig. 2. The volume of net export potential of the Russian Federation in 2016 was estimated at 658 million tons of oil equivalent, accounting for about 5% of the total global primary energy consumption. Over the last 10 years a net export potential of Russia increased by 19% (from the volume in 2005), and its share in relation to the volume produced in Russia energy resources for the same period increased from 46 to 50% (Ivanov and Matveev 2017).

In the group of the largest net importers leading role belongs to the countries of South-East Asia (China, Japan, India). Absolute leader among them is China. The second among them in amount of energy consumption is Japan. In this country, through the measures to improve energy efficiency, the fuel consumption will not increase, but the share of imported energy in total consumption in recent years has increased due to the closure of nuclear power plants after the accident at the AES « Fukushima-1 » . After Japan, followed by India growing industry of which encourages the country to intensify the import of energy resources.

Describing the market position of the largest agents of global energy markets should be noted the rapid changes over the last 10 years in the position of the United States. Taking a course on self-sufficiency with fuel and energy resources due to the extraction of shale oil and gas, the country has reduced the share of imported fuels in the domestic energy mix from 31 to 10% and moved from the first to the fifth place among the net importers of energy (Ivanov and Matveev 2017).

6   The Transformation of the Market Environment Over the Last 5–10 Years

The main long-term trend is the reduction of energy intensity (specific energy consumption per unit of gross product) at national and global levels. The latest short —term and medium—term trends of the transformation of energy markets are closely connected with the specifics of global economic development (in general). Its specific is: the slowdown of the global economy; growth of the weight of Asia in global gross product; rapid increase of differentiation of countries by GDP growth rate; the slowdown of China's economic development; crises in Brazil, Russia and South Africa. In China, country with the world's largest economy and the largest volume of demand imposed on the global energy market is witnessing radical changes in the energy sector in order to reduce the energy intensity of the economy and solutions to environmental problems (Makarov et al. 2016). In this regard, the rate of growth of total world energy demand is decreasing.

Every year it is possible more and more economically conserve the fuel per unit of gross domestic product. So, from 2011 to 2015 global GDP grew by 5% (average over 1 year), while global energy consumption increased by only 1.5% (average over 1 year). The most developed countries in the OECD in 2007 had reached the maximum of the absolute amount of average annual consumption of energy resources, after the volume was reduced (Ivanov and Matveev 2017). However, in general, in the world took place the expansion of energy consumption (over the last decade at 20.2%). It grew rapidly due to developing countries, primarily China, Brazil and India.

Over the past 5 years in the background of slowdown in the absolute volume of energy consumption is accelerating the pace of production, first of all, it is based on hydrocarbons. So, the United States through the introduction of new technologies for the extraction of shale oil and gas has reduced the share of imported energy goods doubled from 23.4 to 12.6% of the total requirements of this country. In early 2015, the United States was in first place in the world in daily oil production −13975 thousand barrels\day. (against 11624 in Saudi Arabia and 10853 in Russia) (IEA 2015). On the market of hydrocarbon energy left Iran, Iraq, and Brazil. The oil reserves in the vaults of OECD countries reached an unprecedented high level.

This has resulted an increase in the total supply and reducing demand in the international energy market. Global production of energy resources increased by 9.9% and consumption by 7.9%. The reduction of the growth rate in energy demand while increasing the growth rate of their proposals led to the fact that the production exceeded the volume of consumption and the global energy market from a seller's market has transformed into buyers market (Fig. 3).

«Revolution » in the ratio of supply and demand on the global energy market cased an enormous reduction of prices. The decrease of oil prices in 2014–2016 illustrated in Fig. 4. Since prices of oil, gas and other hydrocarbon energy resources are closely interrelated, Fig. 4 shows the common trend.

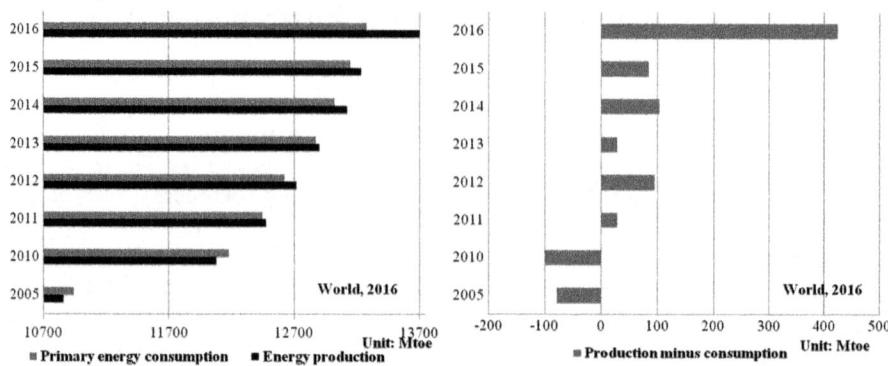

Fig. 3 (World energy production and World Primary energy consumption 2005–2016, Million tonnes oil equivalent (Mtoe))

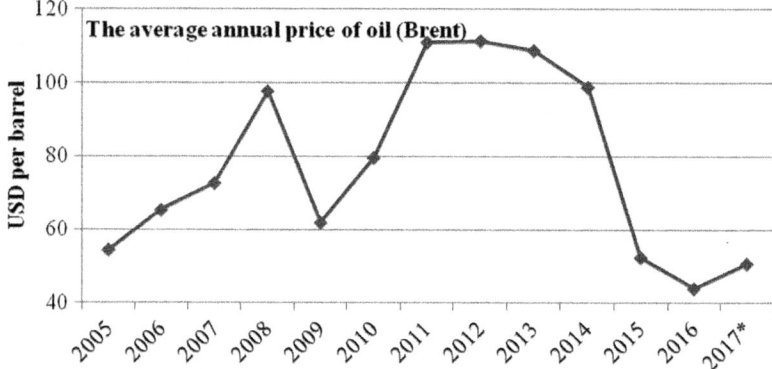

Fig. 4 The average annual price of oil (Brent), USD per barrel *Sources* Ereport.ru; Bloomberg.com

Illustrated in Fig. 4 fall in oil prices was also a result of dumping of the largest oil producers (OPEC, US and Russia), their desire to capture the market. For example, in 2015–2016 Saudi Arabia introduced a discount from the market price of oil exported to Asian and European markets (Slav 2016; Faucon 2016). In the United States and Canada cases of sale of oil at negative cost have documented, when the producers of petroleum products was cheaper to pay the buyers for the export of oil than to keep it as non-liquid goods (RIA Nakanune.ru 2016).

Several authors puts forward the assumption that not only excess volume of supply over demand in the world market, the collapse of oil prices in 2014 could be caused by direct or indirect manipulation of oil prices by individual countries. For example, a decrease in the price of oil might be interested importers of hydrocarbon

Table 1 The structure of world primary energy consumption by types of fuel and energy resources in 2005, 2010, 2014–2016 (%)

Indicator/Year	2005	2010	2014	2015	2016
Primary energy	100	100	100	100	100
Oil	36	33.5	32.6	32.9	33.2
Gas	23	23.7	23.7	23.8	24.1
Coal	28.6	29.8	30	29.2	28.1
Nuclear energy	5.7	5.2	4.4	4.4	4.5
Hydroelectricity	6	6.4	6.9	6.8	6.9
Renewables	0,7	1.4	2.4	2.9	3.2

Sourse VR 2017

fuels (Husain et al. 2015), as well as oil producers with low costs of production displace new suppliers with presumably higher cost (Faucon et al. 2016).

Dynamic changes in world energy markets have contributed the deepening of the economic and political contradictions between Russia and the West, which reinforced its overall geopolitical tensions. In these circumstances, more and more countries, including developing countries, preparing measures to reduce dependence on imports of hydrocarbons. Structure of global energy balance changing, which increases the proportion of renewable energy sources (RES)—solar, wind, tidal, biofuels energy etc. by reducing the share of nuclear energy in the relative stability of the share of oil and gas and some fluctuations in the share of coal (Table 1).

In general, the main direction of transformation of the global energy market is because the fact that instead of resource globalization comes the pursuit of high EE at all levels of the global economy, to achieve regional sufficiency of own energy resources, wide use of innovative technologies to development of unconventional resources (Inshakova et al. 2016).

7 The Main Factors and Prospects of the Global Energy Markets Transformation

Transformation of energy markets occurs under the influence of the following major interrelated factors (Energy Strategy of Russia for the period until 2035): the development of new technologies of production and consumption of energy resources; increase of the competitiveness of renewable energy. And also, the policy of the countries—major consumers and suppliers of energy products, including in relation to the regulation of global energy markets.

New technologies plays a key role in the development of the industry and rank energy. Despite the fact that according to the forecast of the UN experts fossil fuel will come on the basis of the global energy mix the plot up to the end of the XXI century, the development of the country develop a strategy for global development based on three « E »: energy, economy and ecology (Ivanov and Matveyev 2017).

To such technologies, which eventually lead to a course during a state of hydro-carbons in the global energy markets include: renewable energy sources and storage devices; electric vehicles, wood fuel cars; web technologies (active-adaptive net-work, the distribution of production and other);—energy-efficient technologies in the construction sector (zero-energy buildings, « smart home » , « smart city »).

For countries with developed market economy, new energy-efficient technolo-gies are the main factor of changes in demand for energy, while for developing countries demand mainly depends on population growth and economic growth. In the long term perspective the development of energy efficient technologies and technologies of processing have the potential to significant reduction of energy consumption and to the closed cycle of energy exchange (Bushuev et al. 2016, p.8).

The most important direction of development of new energy technologies is the development of electricity generation based on renewable energy sources (RES)—solar and wind energy, and biomass. Prospects of renewable energy customary to consider in assessing the natural potential, but also the economic benefits (or incentives). At present, almost ubiquitous production of electricity based on wind power and solar is more expensive than generation based on traditional hydrocar-bon energy resources. However, as more technological progress with each doubling of electricity production based on renewable energy, cost of the product is reduced by 20% (IEA 2017). Continuation if this trend can ensure the feasibility of the widespread transition to renewables in the medium term. Today energy capacities creation based on renewable energy cost-effective in remote places where it's expensive to hold on the power lines, which is very important for the Russian areas in the North and far East.

Russia, despite its vast hydrocarbon reserves, is a supporter of renewable energy and consistently implements the state policy in this sphere. In 2017 within the framework of the international forum « Russian energy week » , Deputy Minister of Energy Alexei Teksler said that Russia, as one of the leaders in global energy is an active participant in its global « green » transformation. However, he stressed that sustainable energy development should include traditional energy to be inte-grated, balanced and consistent with the interests of different groups of countries. It is advisable to consider the use of the most environmentally friendly of fossil fuels, primarily—natural gas, and also to pay attention to improving the environmental sustainability of fossil fuels in cases, where the rejection of them is impossible. Accordingly, the fuel and energy balance of Russia today are very environmentally friendly: more than half of primary energy consumption is gas, and in electricity production on carbon-free or low-carbon sources have altogether more than 80% (Teksler 2017).

Finally, the decisive subjective factor in the development of the global energy markets are targeted measures to diversify the energy balances of most countries. These measures are developed and implemented by governments in the framework of the energy component pursued their economic policies. They aimed at diversi-fication of suppliers of hydrocarbon raw materials, expanding the use of renewable energy sources and local fuels, the transformation of the regulation of global energy markets, including changes in the terms of contracts and the evolution of the regulation of exchanges to strengthen the position of consumers.

However, according to most forecasts, including the ERI RAS forecast, in future perspective up to 2035, the global energy will continue to be primarily carbon with a hydrocarbons dominance (Makarov et al. 2016). Despite an active development of distributed generation at the heart of electricity of most countries in the world in the run up to 2035 will be the existing system of centralized electricity supply. It will be based on large power plants, namely: TES, NPS, HES or renewable energy sources —the so-called « network » wind and solar power plant. Therefore, there is many reasons to expect a continuation technological development of traditional energy with the steady increase of their efficiency and environmental performance.

It is expected, the European market, which is key market for Russia, the energy-efficiency consumption or even decrease slightly due to energy efficiency, however, due to falling domestic production, the European countries will be forced to increase imports of fossil fuels (ERI RAS forecast). In this, energy policy of the EU aimed at diversifying supply sources and increasing the share of renewable energy in the energy continued to grow and were greater momentum. In the near future, the main growth in energy demand expected in the Asia-Pacific region (APR), and then in Africa. All this opens up new opportunities for Russian energy, but requires large investments in the development of appropriate transport infrastructure. Large pipeline and infrastructure projects require non-trivial solutions in terms of financial support for their construction. In this regard, the mechanism of syndicated lending to the project company, created by the banks participating in the syndicate, that controls the subsequent use at all stages and the volume of mastering of credit funds (Inshakova et al. 2017) is effective.

8 Energy Markets as Objects of Geopolitics and Global Energy Security

The geopolitical aspect of the analysis of the development of world energy markets involves a considering what is happening with them changes as managed processes developing under the action of purposeful measures of the state policy of the countries-leaders of economic development of the countries-major consumers and producers of primary energy role in international relations. However, in the context of global economic growth and increasing demand for energy (on one side) and nerves spatial distribution of geological spas of hydrocarbons as tradition primary energy source (the other side)—the dependence of most countries economic from outside the post-war energy became problematic. Since the middle of 1970s, energy is increasingly becoming a subject of geopolitics of different countries and their associations, and its regulation—a cause of political tensions and inter-state conflicts (Shafranik 2015, p.13).

The beginning of the XXI century marked by the increase in demand on tradition energy resources, the high growth rate of oil, gas and electricity prices, as well as the expectation exhaustion of operated hydrocarbons in the near future. Against this background occurred the « wet revolutions » , wars in oil-producing Arab countries associated with the impact forces weep.

In the period of sharp growth of prices for hydrocarbon raw material, the main energy consumers on the world market—US, EU and China—have developed state programs. These programmes related to reduce dependence of their economic importance from energy import, enhance their own energy safety and ranches geopolitical positions of the countries-exporters of hydrocarbon resources, including Russia, on the global markets. (Belogoryev et al. 2011).

The rapid growth of demand and oil prices that took place in the 2000s, stimulated shale revolution and ended in the mid-2010s with the advent of new exporting countries of oil and gas. As a result, by 2014 on the world oil market, the offer has exceeded demand, and prices of oil, natural gas and other energy sources has dramatically decreased (Fig. 4). International competition for markets of hydrocarbons, which is exacerbated in light of the lifting of international sanctions on Iran and willingness to pay any price to maintain its share of the world oil market. All this had a restrictive impact on the export of Russian oil has increased.

In 2014, the extremely acute social and military conflicts in Ukraine and the middle East. In 2016, began a massive migration of refugees to Europe; intensified international terrorism and expanded its area.

In 2014–2017 in connection with Ukraine events the United States of America, the European Union and several other countries imposed economic sanctions against Russia. In particular, in 2014 from the US and the EU introduced sectoral sanctions against the major oil companies in Russia: ACS « Rosneft » , WAP « Gazprom » , WAP « LUKOIL » , ACS « Surgutneftegaz » , ACS « Gazprom Neft » , ACS « NOVATEK » and ACS « Transneft » . These sanctions limit the supply to Russia of equipment and technologies for the development of the oil industry on the basis of the development of hydrocarbon resources of the deep shelf exploration and production of hydrocarbons in the Arctic, and shale oil mining.

All this exacerbates the geopolitical contradictions between two countries, leads to changes in the flows of energy supply, represents the beginning of a redistribution of the market between the major parties. A big destabilizing influence on the global economic system and its energy subsystem had « information war » with anti-Russian orientation.

In this situation changes the perception of the global energy security. Traditionally the electrical safety means such conditions under which consumer has reliable access to energy, and a supplier to its consumers (Sapir 2006). In a global aspect in this definition a few years ago, the emphasis was on the protection of the interests of the countries-importers of energy resources. However, in modern conditions in the context of global energy security aspect of ensuring sustainability of the world's energy markets has been mainstreamed. According to the definition given in the Energy strategy of Russia for the period up to 2030: « Energy security

is security state of the country, its citizens, society, state and economy from threats
to reliable fuel and energy resources».[1] It supposed that threats are determined by
external factors—geopolitical, macroeconomic and market, as well as sustainable
functioning of the markets of the energy sector of global and national economy. In
such conditions, the more necessary is the stability of property relations governed
by Russian national legislation (Inshakova et al. 2017).

9 Problems of Sustainable Development Security of World Energy Markets and Their Solutions Based on International Cooperation

Threats of global energetic security related to the fact that the modern world is
experiencing a global transformation period, arising from the redistribution of the
world's potential strength and development, its displacement in Asian-Pacific
region. Contradictions associated with the uneven world development, deepening
gap between the level of world development, an increase in leadership in resource
competition, access to the market, the control over the transport infrastructure,
which primarily relates to energy, energy markets and energetic infrastructure were
growing. (Foreign policy concept of the Russian Federation 2016).

Given the current geopolitical situation in the foreign policy Concept of the
Russian Federation proclaims that « Russia attaches a great importance to sustainable manageability ensuring of global development, which requires collective
leadership by leading States… For these purposes, Russia is stepping up cooperation with partners in the framework of the « group of twenty » , BRICS (Brazil,
Russia, India, China, South Africa), SCO (Shanghai cooperation organization), RIC
(Russia, India, China), as well as in other structures and dialogue platforms » . (The
conception of foreign policy of the Russian Federation 2016, R. 25)

Russia as a major energy power can play an important role in ensuring global
energy security through sustainable development of world and national energy
markets. To achieve this goal in the draft Energy strategy of Russia for the period
up to 2035 (Energy Strategy of Russia for the period till 2035), developed by the
Ministry of energy of the Russian Federation, formulated a number of problems that
will be solved in the field of international relations. These tasks include diversification of destinations and product structure of Russian energy exports; the elimination of discrimination of Russian energy goods, services and investments in the
global energy markets; the convergence of geopolitical and economic interests of
Russia with the respective interests of the major participants in the global energy
markets coordinating with them for ongoing energy policy measures. The formation
of common energy market of the Eurasian economic Union (crude oil, petroleum
products, natural gas and electrical energy) is scheduled. Assumes implementation

---

[1]Energy Strategy of Russia for the period till 2030.

of the General principles of their regulation, including the coordinated policy in the sphere of subsoil use and investments in the energy sector; ensure a free movement of energy products and energy technologies.

To solve these problems, planned the following measures external energy policy: the expansion of energy cooperation with the countries-participants of the EEC, CIS, EU, Shanghai Cooperation Organization, BRICS, the Association of South-East Asia, the East Asian economic community forum « Asia-Pacific economic cooperation » , « Economic and Social Commission for Asia and the Pacific » (ESCAP). And also with countries of the black sea, the Caspian and Arctic regions, North and Latin America, with other international organizations and international multilateral entities; active participation in international negotiations on energy issues.

Due to the need for solve the problems of Russia's energy policy a lot of work on development of normative-legal base of international cooperation gas been mainstreamed. Its include: the consolidation in international law the principle of balance of exporters, importers and transistors interests of energy resources; legal provision of favorable tax, tariffs and customs conditions for export diversification. And also, the legal protection of Russian investments; active participation in the development of international cooperation in the field of environmental security and climate change on the planet; international-legal registration of external border of continental shelf of the Russian Federation in the Arctic ocean; and other aspects.

10   Conclusion

The modern condition of the global energy industry is characterized by a change in technological structure and the worsening of geopolitical contradictions between the countries-producers and consumers of energy resources. The largest net exporter of energy resources in the world market are Russia, Saudi Arabia, Australia, Indonesia, Norway and Kuwait, and among the world's net importers leaders are China and other countries of South-East Asia.

The transformation of the global energy markets manifests itself in such processes and phenomena as reduction of the energy intensity of the global and national economies; increasing the share of non-carbon and low-carbon energy in the structure of the world balance of primary energy. And also, redistribution of monopoly market power from exporting countries to importing countries hydrocarbons; a high degree of uncertainty and risk for all market participants, volatility in energy prices. To replace a resource of globalization comes the desire to achieve regional sufficiency of own energy resources.

The main driving forces of the transformation of the global energy markets are: the development of new technologies of production and consumption of energy resources; improving the competitiveness of renewable energy sources; divergent geopolitical interests of the countries—consumers and suppliers of energy products. The aggravation of contradictions between the countries—participants of the energy markets leads to changes in the flows of energy supply, the redistribution of markets between the main participants.

The external energy policy of Russia is aimed at ensuring global energy security through sustainable development of world energy markets. To achieve this goal will require broad international cooperation in the solution of such tasks as diversification of Russian energy exports; the elimination of discrimination of Russian energy companies; approval of Russia's interests with the interests of other participants in the energy markets. All of this aggravated a lot of work in the field of international law on the legal framework of this cooperation and the regulation of global energy markets.

References

Belogoryev, A. M., Bushuev, V. V., Gromov, A. I., Kurichev, N. K., Mastepanov, A. M. & Troitskiy, A. A. (2011). Trends and scenarios of world energy development in the first half of the 21st century, Energiya Publishing House, Moscow. Retrieved September 20, 2017, from http://www.energystrategy.ru/editions/trends.htm. (Russian).

Bloomberg.com. Retrieved October 20, 2017, from https://www.bloomberg.com/energy.

BP. (2016). Statistical Review of World Energy, June 2016. Retrieved September 20, 2017, from https://www.bp.com/content/dam/bp/pdf/energy-economics/statistical-review-2016/bp-statistical-review-of-world-energy-2016-full-report.pdf.

BP. (2017). Statistical Review of World Energy, June 2017. Retrieved September 20, 2017, from https://www.bp.com/content/dam/bp/en/corporate/pdf/energy-economics/statistical-review-2017/bp-statistical-review-of-world-energy-2017-full-report.pdf.

Bushuev, V. V., Gromov, A. I., Belogoryev, A. M., & Mastepanov (2016). Power industry of Russia: A post-strategic look for 50 years ahead, Energiya Publishing House, Moscow. Retrieved September 20, 2017, from http://www.energystrategy.ru/editions/docs/energy_Russia_50.pdf p. 8, (Russian).

Enerdata. (2017). Enerdata intelligence + consulting, Global Energy Statistical Yearbook 2017. Retrieved September 25, 2017, from https://yearbook.enerdata.net/total-energy/world-energy-intensity-gdp-data.html.

Energy Strategy of Russia for the period till 2030, (approved by Decree N°1715-r of the Government of the Russian Federation dated 13 November 2009), Moscow 2010. Retrieved October 20, 2017, from https://minenergo.gov.ru/node/1026 (Russian).

Energy Strategy of Russia for the period till 2035, (project), Ministry of Energy of the Russian Federation. Retrieved October 20, 2017, from https://minenergo.gov.ru/system/download-pdf/1920/69055. (Russian).

Ereport.ru, World Economy. (2017). Retrieved September 25, 2017, from http://www.ereport.ru.

Faucon, B., & Said, S., Chaturvedi, S. (2016). Saudi Arabia cuts Asian oil prices to counter rivals Russia, Iraq and Iran. Wall Street Journal, Aug. 5, 2016. Retrieved October 20, 2017, from http://www.wsj.com/articles/saudi-arabia-cuts-asian-oil-prices-to-counter-rivals-russia-iraq-and-iran-1470416304.

Faucon, B. (2016). Saudi Arabia cuts oil prices in Europe as Iran ramps up exports. *Wall Street Journal*, Jun 05. Retrieved October 20, 2017, from 2016, https://www.wsj.com/articles/saudi-arabia-cuts-oil-prices-in-europe-as-iran-ramps-up-exports-1465165449.

Husain, A. M. et al. (2015). Global implications of lower oil prices. IMF. (2015). Retrieved October 20, 2017, from https://www.imf.org/external/pubs/ft/sdn/2015/sdn1515.pdf.

IEA. (2017). International Energy Agency, Energy Efficiency 2017. Retrieved September 20, 2017, from https://www.iea.org/publications/freepublications/publication/Energy_Efficiency_ 2017.pdf.

Inshakova, A. O., Frolov, D. P., Kazachenok, S. Y., & Maruschak, I. (2016). Institutionalization of Intellectual property on resource-saving technologies and materials: A comparative institutional study of usa and Russia. *Journal of Advanced Research in Law and Economics, 6*(20), 1373–1382.

Inshakova, A. O., Goncharov, A. I., & Sevostyanov, M. V. (2017). Institutional ambiguity of regulation of possessory relations in modern Russia. *Overcoming Uncertainty of Institutional Environment as a Tool of Global Crisis Management, 1,* 207–212.

Inshakova, A. O., Goncharov, A. I., Kazachenok, O. P., & Kochetkova, S. Y. (2017b). Syndicated lending: intensification of transactions and development of legal regulation in modern Russia. *Journal of Advanced Research in Law and Economics, 3*(25), 838–842.

Ivanov, A. S., & Matveev, I. E. (2017). The world energy market under geopolitical realities on the eve of 2017. *Russian Foreign Economic Journal, 1,* 17–31. (Russian).

Makarov, A. A., Grigoryev, L. M., & Mitrova, T. A. (eds.) (2016). Forecast of the development of the world's and Russia's energy—2016, Institute of Energy Research, RAS & Analytical Centre at the Government of the Russian Federation, Moscow. Retrieved September 1, 2017, from http://ac.gov.ru/files/publication/a/10585.pdf. (Russian).

RIA Nakanune.RU. (2016). The oil price from Northern Dakota fell below zero—from sellers demand to pay in addition to the buyer. Russian information agency Nakanune.RU, Jan 18, 2016. Retrieved October 20, 2017, from http://www.nakanune.ru/news/2016/01/18/22425242 (Russian).

Sapir, J. (2006). Energy security as general benefit. *Russia in Global Policy, 6* (2006). Retrieved September 20, 2017, from http://www.globalaffairs.ru/number/n_7780 (Russian).

Shafranik, Yu. K. (ed.) (2015). Global energy and geopolitics (russia and the world). Energiya Publishing House, Moscow. Retrieved September 1, 2017, from http://www.energystrategy.ru/ editions/docs/globalenergy.pdf.

Slav, I. (2016). Saudi arabia slashes crude price to Asia. Oilprice.com, Aug 01, 2016. Retrieved October 20, 2017, from http://oilprice.com/Energy/Energy-General/Saudi-Arabia-SlashesCrude-Price-To-Asia.html.

Teksler, A. (2017). Alexey Teksler on "The Russian power week": The renewable power in Russia shows quantitative and high-quality growth. *Ministry of Energy of the Russian Federation.* Retrieved November 20, 2017, from https://minenergo.gov.ru/node/9455, (Russian).

The Conception of Foreign Policy of the Russian Federation. (2016). (approved by the Decree of the President of the Russian Federation V. V. Putin of November 30, 2016). Retrieved November 10, 2017, from http://www.mid.ru/foreign_policy/news/-/asset_publisher/ cKNonkJE02Bw/content/id/2542248 (Russian).

WEC. (2016). World Energy Concil, World Energy Resources 2016. Retrieved November 20, 2017, from https://www.worldenergy.org/wp-content/uploads/2016/10/World-Energy-Resources_Full.

Part III
Authority of International Legal Entities and Integration Unions on the Development of Civil Legal Regulation of the Foreign Trade Turnover of Energy Resources

TNCs as Subjects of Economic Activity and Lawmaking in the Sphere of Foreign Trade in Energy Resources

Agnessa O. Inshakova, Evgenia E. Frolova and Igor P. Marchukov

1 Annotation

The chapter outlines the range of subjects, and identifies possible legal forms of participation in the foreign trade turnover of energy resources in accordance with the current Russian legislation, among which the main place belongs to TNCs. Based on the analysis of the doctrines presented in the doctrine and international lawmaking, the specific features of the legal status of TNCs, as well as characterizing this form of management signs that determine their prevailing participation in the field of foreign trade energy relations. The participation of TNCs as the main subjects of foreign trade in energy resources and their significant impact on its legal regulation is also justified through the study of a number of legal documents, officially published statistical data, as well as materials of international judicial practice of proceedings between the largest energy companies—TNCs of Russian nationality. The authors come to the conclusion that TNCs are not simply the main (primary) subjects of the legal relationships under investigation, but at the same time, they represent a powerful mechanism for influencing and regulating them.

Based on the changes introduced into the Civil Code of the Russian Federation reflecting the most modern tendencies in the development of conflict-of-law

A. O. Inshakova (✉)
Department of Civil and International Private Law, Institute of Law,
Volgograd State University, Moscow, Russia
e-mail: gimchp@volsu.ru; ainshakova@list.ru

E. E. Frolova
Institute of State and Law, Russian Academy of Sciences, Volgograd, Russia
e-mail: frolova_ee@rudn.university

I. P. Marchukov
Department of Civil and International Private Law, Volgograd State University,
Volgograd, Russia
e-mail: gimchp@volsu.ru

© Springer International Publishing AG, part of Springer Nature 2019
O. V. Inshakov et al. (eds.), *Energy Sector: A Systemic Analysis of Economy,
Foreign Trade and Legal Regulations*, Lecture Notes in Networks and Systems 44,
https://doi.org/10.1007/978-3-319-90966-0_11

regulation, it is concluded that an important role in the process of such development in the sphere of civil regulation of the foreign trade turnover of energy resources is assigned to various kinds of international legal entities, organizations and integration associations.

This, in turn, is caused by the expansion of international energy cooperation in foreign economic activity. Perspective tendencies of development of conflict regulation of foreign trade turnover of energy resources are determined.

It is alleged that through the activities of TNCs, the integration processes of development and liberalization of energy regulation are intensified, as a result of which the role and the foundations of international private contractual practice—the international commercial law of legal entities—energy companies, which acquires increasing importance in the general legal array and should be taken into account in the process of lawmaking.

A particularly significant contribution to the unification of the interests of states at the level of economic entities, the development and creation of a uniform private-law regulation of foreign trade relations in the energy sector in the context of expanding international cooperation of the Russian Federation, international integration associations with the participation of the Russian Federation and (or) its main foreign economic partners are contributing.

2 Materials

Legal relations in the field of energy between persons engaged in the search, extraction, production, processing, sale, storage, distribution of various types of energy resources, engineering surveys, design, construction, modernization, reconstruction of power facilities, as well as between persons engaged in innovative activities and other activities in the energy sector, and persons who acquire various types of energy resources and who are provided with relevant services are studied in the chapter, first of all, on the normative basis of a number of federal laws of the Russian Federation. Among them: the Civil Code of the Russian Federation, Federal Law No. 164-FZ of 08.12.2003 "On the Basics of State Regulation of Foreign Trade Activity", Federal Law No. 317-FZ of 01.12.2007 (as amended on 03.07.2016) "On the State Corporation for Atomic energy "Rosatom"", Federal Law No. 115-FZ of 21.07.2005 "On Concession Agreements", Federal Law No. 39-FZ of 25.02.1999 "On Investment Activities in the Russian Federation in the Form of Capital Investments".

The legal policy of the activities of business entities operating in the energy sector was studied on the basis of international, program and strategic acts of the Russian Federation, such as: The Tripartite Declaration of Principles Concerning Multinational Enterprises and Social Policy, The agreement of the EEA member countries by 2025 in the framework of the Union for the creation of common markets for energy resources (electricity, oil, oil products and gas) (the Eurasian Economic Union Treaty), Order of the Government of the Russian Federation

No. 1715-r of 13.11.2009 "On the Energy Strategy of Russia for the period until 2030", Order of the Government of the Russian Federation No. 1662-r of 17.11.2008 "On the Concept of Long-Term Social and Economic Development of the Russian Federation for the Period to 2020", Decree of the Government of the Russian Federation No. 321 of 15.04.2014 "On approval of the state program of the Russian Federation" Energy Efficiency and Energy Development ", Forecast of scientific and technological development of the fuel and energy sector of Russia for the period up to 2035, Decree of the President of the Russian Federation of 30.11.2016 No. 640 "On the approval of the Foreign Policy Concept of the Russian Federation".

The statistical basis of the study was formed by referring to the official websites of the largest Russian oil and gas companies in the Russian Federation that carry out cross-border management (PAO Rosatom, PJSC Gazprom, etc.).

The empirical data of the study were obtained due to the analysis of the practical activities of the Russian (PJSC Rosneft, PJSC LUKOIL, Surgutneftegaz, PJSC Gazprom Neft) and foreign (Exxon Mobil, Chevron Corporation, CNPC, BP, Shell, Total) oil and gas holdings of the world.

The role of Russian transnational corporations as participants in the energy sector of the economy, as well as their place in the system of international legal regulation of energy markets, was studied using the example of Doronina and Semilyutina (2012), Inshakova (2012), Shiyanova (2016) and others.

Theoretical approaches to the definition of the concept, legal personality and characteristics of TNCs, as well as their normative consolidation, were the works of Asoskov (2000), Boguslavsky (2010), Velyaminov (2015), Inshakova 2012, Calderon (1984), Lunts (1976), Natalukha (1985), Schmitthoff (1973).

The organizational and legal basis of TNCs and the legitimacy of their presence in international energy markets, including as vertically integrated oil companies, were studied thanks to the work of Kozlov (2014), Krivorotko (2015), Chuku (2013), Shumilov (2003).

The issues of influence of international legal entities on the development of legal regulation of foreign trade in energy resources were studied with the help of Merezhko (2012), Vilkova (2002).

The jurisprudence of resolving international commercial disputes between TNCs in the field of energy was investigated on the basis of the following incidents: the case of the Philippines in international arbitration on the basis of a claim against China for territories in the South China Sea in accordance with the UN Convention on the Law of the Sea; dispute between Argentina and the Spanish oil company Repsol about the nationalization of assets; the dispute between the United States and the British company BP in connection with the accident at Deepwater Horizon deepwater drilling rig in the Gulf of Mexico in April 2010 and the consequent technogenic catastrophe; disputes that arose in 2016 between Gazprom and the Danish company DONG Naturgas A/S, the Polish PGNiG, the Turkish BOTA Petroleum Pipeline Corporation, the Dutch Gas Terra B.V. and Shell Energy

Europe, based in London, on the revision of natural gas prices; claims of Nigerian fishermen and farmers to the British-Dutch company Shell at the High Court of London (March 2012) and to the Royal Court of the Netherlands (October 2012) related to the oil spill in the Ogoniland region in 2008 and the pollution of water bodies.

3 Methods

The methodological basis of the research was a set of general scientific (dialectical, historical, inductive, deductive, analytical, synthetic) and private-science (formal-legal, comparative-legal, interpretational, statistical, process-dynamic) methods.

4 Introduction

Legal relations in the field of energy are formed between persons engaged in the search, extraction, production, processing, sale, storage, distribution of various types of energy resources, engineering surveys, design, construction, modernization, reconstruction of power facilities, as well as those who carry out innovative activities and other activities in the energy sector, and persons who acquire various types of energy resources and who are provided with the appropriate services (Romanova 2014).

The main provisions on subjects of economic activity in the field of foreign trade turnover are fixed in Chap. 3 of Federal Law No. 164-FZ of 08.12.2003 "On the Basics of State Regulation of Foreign Trade Activity" (Federal Law 2003). According to Article 10 of this Law, any Russian or foreign persons have the right to carry out foreign trade activities. This right may be restricted in cases provided for by international treaties of Russia and federal laws.

Consequently, all the participants in foreign trade turnover, including in the energy sector, divide the above-mentioned law into Russian and foreign persons. The first include public legal entities (Russia, its subjects, municipalities), legal entities established in accordance with Russian law and registered on the territory of the Russian Federation, as well as registered as individual entrepreneurs Russian citizens and having in our country a permanent or preferential residence of a stateless person. The remaining participants in foreign trade activities are classified as foreign persons.

5 Subjects of Economic Activity of the Russian Federation in the Sphere of Foreign Trade Turnover of Energy Resources

The Russian Federation and its subjects carry out foreign trade activities through specially created (usually in the form of an enterprise) structures operating in the interests of the state.

As an example of a state-owned enterprise that controls the import and export of energy-related goods and services, one can name the state atomic energy corporation Rosatom. The corporation was established on the basis of Federal Law No. 317-FZ of 01.12.2007 (Federal Law 2007) for the provision of public services and management of state property in the field of nuclear energy use, implementation of sectoral regulatory legal regulation. State corporation Rosatom is the largest generating company in Russia, providing 18.3% of the country's electricity generation. It occupies a leading position in the world market of nuclear technologies: 1 place in the world in terms of the number of nuclear power plants being built simultaneously abroad; 2 place in the world for uranium reserves and 4 place in terms of its production; The 2nd place in the world for the generation of nuclear energy, providing 36% of the world market of uranium enrichment services and 17% of the nuclear fuel market (Rosatom 2017). Rosatom organizes the implementation of various activities, including the production, transmission, distribution and sale of electricity and heat energy, and also rendering of the services connected with it (performance of works); export and import of goods (works, services) related to the use of atomic energy; direct investment abroad.

At the same time, the current legislation limits the possibility of direct participation of public and legal entities in foreign trade activities. So, the legislator points out that its implementation is possible only in cases established by federal laws. As an example, we can refer to the Federal Law of 21.07.2005 No. 115-FZ "On Concession Agreements" (Federal Law 2005), in accordance with which, as a concessor, having the opportunity to conclude a contract with a foreign legal entity, public legal entities act on behalf of and in the interests of which the relevant authorities operate.

Unlike production sharing agreements, concession agreements involve granting foreign investors for a certain period of time the right to develop and develop renewable and non-renewable natural resources, as well as entrepreneurial activities related to the use of objects in state ownership. The possibility of concluding a concession agreement with Russian and foreign investors is enshrined in Article 11 of Federal Law No. 39-FZ of 25.02.1999 "On Investment Activities in the Russian Federation in the Form of Capital Investments" (Federal Law 1998).

6 The Role of Transnational Corporations in the Energy Sector of the Economy in the Context of the Globalization of the Economy

Along with the considered legal forms, the role of transnational corporations is now growing, which due to the inherent characteristics of their legal nature in the greatest extent meet the interests of internationalization and concentration of capital at the present stage. The fact of shifting the model of competition from national economies to other global economic factors between economic actors in the energy sector is obvious. Increasing priority in the energy sector is given to TNCs—transnational corporations, supranational entities, interstate economic unions, which is a consequence of the main characteristic features of the globalization process as an objective phenomenon of the economic structure.

In the context of the globalization of the world community, accompanied by the expansion of cross-border economic interaction and interdependence of states, a significant role is assigned to international companies as direct actors of the microeconomic level of modern economic activity. The importance of the role assigned to them in the energy sector is due to the ever growing concentration of their control over strategically important spheres of economic activity and the factors that ensure its production—capital, labor, technology, supplies of raw materials and components, services and sales (Inshakova 2012).

At present time, the topic of transnationalization in the world market of energy resources is quite urgent. Russia has embarked on a course of full-scale integration into the world economy, an important part of which is energy (Order of the Government of the Russian Federation 2008, 2009; Decree of the Government of the Russian Federation 2014; Eurasian Economic Commission 2017). The development of energy cooperation with leading importers and exporters of energy resources is an important element in the implementation of this strategy (Order of the Government of the Russian Federation 2009; Official website of the Ministry of Energy of the Russian Federation 2017; Presidential Decree 2016).

A significant object of international relations, regulated internationally, both publicly and privately, are high-pressure gas mains, high-voltage systems and transmission lines, main oil pipelines, coal slurry pipelines, oil product pipelines and other stationary facilities intended for the transfer of energy resources. On their construction and operation, industry operators specialize, on the establishment of legal control over which the activities of the largest transnational corporations are directed (Shiyanov 2016).

7 The Concept and Characteristics of a Transnational Corporation: Advantages for the Energy Sector

At present time, the concept of "transnational corporation" does not have an unambiguous legal definition. However, there are competent opinions that it is directly connected with the development of investment legislation and the appeal of developing states to the substantive method of regulation as a way of protecting national interests by the recipient state of foreign capital (Doronina and Semilyutina 2012). In general, the definition is enshrined in a number of international instruments. Thus, according to paragraph 6 of the Tripartite Declaration of Principles concerning Multinational Enterprises and Social Policy (Tripartite Declaration of Principles concerning Multinational Enterprises and Social Policy 2001) of 1977, to multinational corporations it is possible to carry such corporations, under the control or in which property outside the country of their location there are manufacture, service, distribution and other spheres (Inshakova 2008).

In the legal doctrine the concept of TNC is treated differently. So far, the most successful has been the general scientific definition of TNCs, proposed by the well-known British lawyer K. Schmitthoff, which defined TNC as a group of companies with different nationalities connected through holding shares, management control, or by concluding a contract representing economic unity (Schmitthoff 1954).

Following K. Schmitthoff, a number of domestic scientists, defining TNCs, emphasize such an essential characteristic of it as a group personality. Among them, Asoskov (2000), Boguslavsky (1999), Doronina and Semilyutina (2011), Lunts (1976), Kulikov (2004), Osminin (1983), and others.

Thus, L.A. Lunts, identifying TNC, speaks of this multinational enterprise as an economic entity with legal multiplicity (Lunts 1976).

M.M. Boguslavsky also believes that the registration of economic unity in TNC through legal multiplicity, serves the interests of the owners of the enterprise (Boguslavsky 2010).

B.I. Osminin also sees the peculiarity of TNC as an object of legal regulation in that it is a powerful parent company that leads a wide network of its foreign enterprises and organizationally and economically forms a single entity that is combined with legal multiplicity (Osminin 1983).

Defining the international legal nature of TNC, N.G. Doronina and N.G. Semilyutina drew attention to the fact that "international companies include multinational enterprises that have an extensive network of branches and subsidiaries throughout the world. At the same time, economic subordination and legal independence of subsidiaries from the parent organization are the main characteristics of such international organizations "(Doronina and Semilyutina 2011).

Despite the fact that as the main characteristic of the legal nature of the TNC sign A.V. Asoskov singles out centralized (unified) management of the activities of TNC, nevertheless, the scientific works of the author clearly show that this feature exists inseparably and on an equal footing "with the presence of several legal entities in the structure of TNC" (Asoskov 2000).

It can be noted that the sign of centralized management in TNC is economic in nature, which is similar to the criterion of exercising control of one company against another from a legal point of view (Shevtsova 2008).

With the correct, in our opinion, positions of perception of TNC as a group of legal entities, the definition of VD Fedchuk is rather detailed, who proposed to define TNCs as "a group of companies that is an economic entity consisting of two or more companies, each of which is an independent legal entity, but which are connected either by the control exercised by one company, called the parent company, over other companies, or by the fact that these companies, being independent, are under general management" (Fedchuk 2010). There is, however, a point of view according to which, in addition to a group of persons, TNCs can also represent a legal entity that has branches and representative offices in other states (Lysenko 2003; Boguslavsky 2010).

As specific features of the TNC, D.L. Lysenko notes that despite the existence of offices in many countries, the enterprise as a whole does not fall under the jurisdiction of any particular country and its activities as a single entity can not be settled by the national law of any state. This is possible only in the framework of international law (Lysenko 2003).

M.I. Kulagin also pointed out that there was a discrepancy between "legal forms and economic essence" (Kulagin 1997), the interest now belongs to one legal entity (the parent company of TNC), and the execution and responsibility to another (controlled entities). The control exercised by the parent company plays a decisive role, replacing the classic internal management of the company by external management (Kulagin 1997).

Velyaminov G.M. proposed to define TNCs as independent of the form of ownership and country of origin of the enterprise, which have companies in two or more countries and function according to the system of decision making, which provides an opportunity to pursue a common strategy and agreed policy (Velyaminov 2015).

The absence of a single generally accepted definition is "complemented" by the absence of specific international norms that can introduce TNCs into the framework of the existing world laworder. Those international acts and agreements, which are in effect for the time being, are mostly recommendatory in nature. International documents that operate with the term "TNC" are either not mandatory for use (for example, the norms relating to the responsibilities of transnational corporations and other human rights enterprises adopted on August 13, 2003 at the 22nd meeting of the Subcommission on the Promotion and Protection of Human Rights of the UN Commission on Human Rights (Scholarship Repository 2017), or are not of a legal nature (for example, the annual World Investment Report (World Investment Report 2001; World Investment Report Transnational Corporations and Export Competitiveness 2002). It means that there are no real legal barriers to the activities of transnational business associations around the world today, it often leads to abuse of the economic power concentrated in their hands.

Attempts to develop internationally a uniform definition of multinationals were undertaken by various international organizations. So, under the auspices of the UN, special structures were created whose activities focused on the integrated study of TNCs (Sharkova et al. 2016). In 1974, on the initiative of the so-called "Group of 77" developing countries, the UN Commission on TNCs was established. In 1993, an appropriate unit of UNCTAD was established to systematically work on the activities of TNCs. In 1975, at a special session of the UN General Assembly, a resolution was adopted, according to which a special commission was set up to develop the Code of Conduct for TNCs (hereinafter referred to as the Code) (Calderon 1984). The draft Code containing the definition of TNC was submitted by 1990, but was never adopted (Natalukha 1985).

In the Draft Code of Conduct for TNCs, a corporation is defined as "an enterprise, regardless of country of origin and form of ownership, including private, state or mixed enterprises, having branches in two or more countries, regardless of the legal form and scope of these offices, which operate in accordance with a specific decision-making system that allows for the implementation of a coherent policy and overall strategy through one or more decision-making centers, and in which the departments are thus interconnected, whether by property relations or other relations, that one or more of them can have a significant impact on the activities of others" (Kulikov 2004).

In the list of the main reasons for the ineffective activities of TNCs within CIS legislation, experts call a wrong approach to the definition of TNCs as a political tool, rather than economic unity (Kamalov 2011). Indeed, neither in the CIS legislation nor in the national legislation of the CIS countries is there a clear mechanism that determines the procedure for the creation and operation of such international corporate structures as TNCs.

Meanwhile, the analysis and gradual solution of such problems of legal regulation of TNCs as issues of uniform definition of TNCs, the relationship between the obligations of TNCs and states, the possibility of granting a national regime to branches of foreign TNCs, the use of so-called customary international law with regard to the activities of TNCs, the rules for the nationalization of TNCs' property and jurisdiction in settling disputes is of great importance for any national legal system. Legislative settlement of these issues is of particular importance for the modernization of modern Russia, which needs strong transnational corporations that promote high-quality economic growth and improve the country's competitiveness in the international arena (Inshakov et al. 2014). Really existing and active in international economic markets, Russian TNCs have long required legislative recognition within the framework of Russian corporate law at the level of the Federal Law "On Transnational Corporations". Within the framework of the Federal Law, it is necessary to determine the legal status of such a corporation, its place in the system of Russian corporate law, to determine the structure of management bodies and their competence, the specifics of the procedure for creating and terminating activities, the specific features of the legal status of TNC participants, the procedure for making transactions (both large and interested-party transactions). The creation of a special regulatory and legal document will allow to streamline the system of Russian corporate regulation, both with respect to Russian and foreign TNCs. And it should begin, of course, with the definition.

The following characteristics are typical for TNCs: economic isolation; a single internal management of the head enterprise, which coordinates the activities of foreign branches; a wide system of trade and economic ties that go beyond the country of registration (Inshakova 2008). These signs and determine the participation of TNCs in the foreign trade turnover of energy resources as the main subjects of economic activity.

Analysis of the definitions presented in the doctrine and international lawmaking allows us to single out those characteristics that are always and inevitably inherent in this form of management as characteristic features of TNCs, which determine their advantages in the sphere of international turnover of energy resources:

- economic integrity by a complex cross-border interconnected ownership relationship or other structure relationships;
- implementation of coordinated activities at the national and international levels;
- legal multiplicity, manifested in the existence of a different legal form of independent and dependent structural units of a legal entity;
- Different forms of ownership of a multinational enterprise;
- centralized management and control of the parent enterprise, ensuring coherence of actions;
- limited responsibility of independent foreign structural divisions (Inshakova 2012).

8 Structure and International Presence of Russian TNCs in the Energy Sector as Holding Entities

It should be noted that the largest Russian multinationals are vertically integrated oil companies (hereinafter referred to as "VINK"), which carry out activities throughout the chain of energy business—from exploration, through extraction and refining, to the marketing of petroleum products to end-users. This strategy is conditioned by the desire of the major participants of the oil and gas market to minimize dependence on other oil corporations that carry out certain activities in the field of transportation, processing or marketing of products, as well as the desire to obtain additional competitive advantages in the market.

These associations of capital and production in a vertically integrated structure represent complex holding entities, the concentration of management and control (Parfenov 1999). The organizational-legal basis of VINK is a joint-stock company of the holding type, which is the main company in relation to other subsidiaries of oil refineries and other companies, and manages cash flow, production, coordination and control. At the same time, subsidiaries are almost independent of VINK and have their own management. The main society entrusts the execution of specific tasks that contribute to the achievement of certain targets.

The largest VINK—oil companies include Rosneft, LUKOIL, Surgutneftegaz, and PJSC Gazprom Neft—about 80% of the market (Kozlov 2014). For foreign companies the business model includes all major oil and gas companies of the world: Exxon Mobil, Chevron Corporation, CNPC, BP, Shell, Total and others.

One of the largest Russian gas TNCs is PJSC Gazprom, the largest shareholder of which is the Russian state (38% of shares). The revenues from the export of this company form ¼ of the budget revenues of Russia and almost 20% of all foreign exchange earnings (Krivorotko 2015).

The main activities of this global energy company are the extraction and sale of gas, oil, production and sales of electricity and heat energy.

Gazprom has the world's richest natural gas reserves. Its share in world gas reserves is 17%, in Russian—72%. The company is a gas supplier to both Russian and foreign customers and has the world's largest gas transmission system, which is 171,200 km long.

In the domestic market, Gazprom sells more than half of the gas sold. In addition, the company supplies gas to more than 30 countries in the near and far abroad (Gazprom 2017a, b).

The largest foreign partners of PJSC Gazprom in foreign economic relations (in particular, in the sphere of supply) regulated by international private law are: the Danish DONG Energy; German companies Wintershall Holding, E.ON, Verbundnetz Gas, BASF, Siemens; Chinese CNPC, Petro China; French GDF SUEZ, EDF and Total; Indian GAIL; the Italian ENI; Finnish Fortum; Turkish Botas; Netherlands Gasunie and Gas Terra; Japanese Mitsui, Mitsubishi Corporation and TNC Shell; Vietnamese Petrovietnam, Norwegian Statoil; Austrian OMV; Algerian Sonatrach, Polish PGNiG; the Venezuelan PDVSA; Hungarian MOL; Korean Kogas (Woidasky and Hirth 2012; Gazprom in questions and answers 2017).

Apparently, among the listed counterparties of the Russian corporation Gazprom most of the companies have the nationality of the EU member states, which stipulates the author's appeal to the study of the legal regulation of foreign trade turnover of energy resources in the common law of the EU as an international integration association.

International relations of Gazprom are not exhausted by the above-mentioned companies. In view of the fact that Gazprom is a participant in foreign upstream projects (oil and gas exploration and production), Gazprom International was established, which is its specialized company. The company's foreign partners are: the Ministry of Energy and Industry of the Republic of Tajikistan, the Nigerian National Oil and Gas Corporation, the Uzbekneftegaz NHC, the Petrobangla (Bangladesh Oil, Gas and Mineral Resources Corporation), National Algerian Agency for Hydrocarbon Resources Exploration Alnaft, YPBF (Bolivian National Oil and Gas Company), international network of Deloitte (USA), BAPEX (Bangladesh) and others (Gazprom International 2017).

But on the European gas market, the share of Gazprom is growing continuously and only in 2016 it increased by 3 to 34% (Gazprom 2017b).

Another Russian transnational company, one of the largest vertically integrated oil and gas companies in the world, which accounts for more than 2% of global oil production and about 1% of proven hydrocarbon reserves—LUKOIL. With a full production cycle, the Company fully controls the entire production chain—from oil and gas production to marketing of petroleum products. Every day, the Company's products, energy and heat are bought by millions of consumers in 35 countries of the world (LUKoil 2017).

One of the most powerful domestic oil and gas corporations is Rosneft. The main activities of PJSC NK Rosneft are: search and exploration of hydrocarbon fields, production of oil, gas, gas condensate, implementation of offshore development projects, processing of extracted raw materials, sale of oil, gas and products of their processing in Russia and beyond.

The company is included in the list of strategic enterprises in Russia. Its main shareholder (50.00000001% of shares) is JSC ROSNEFTEGAZ, 100% owned by the state. Rosneft is a global energy company with major assets in Russia and a diversified portfolio in the prospective regions of the international oil and gas business, including assets in Venezuela, the Republic of Ecuador, the Republic of Cuba, Canada, the United States, Brazil, Norway, Germany, Italy, Algeria, Mongolia, China, Vietnam, Turkmenistan, Belarus, Ukraine and the UAE (Official site Rosneft 2017).

The competition of capital in the world market, especially in the energy sector, inevitably leads to the use of TNCs of any means to strengthen the competitive positions of their states. One can see a clear interdependence of national states and their TNCs. The state is interested in developing its own multinationals, which strengthen the competitive positions of their states in the world arena (Chuku 2013).

Thus, in the modern world, TNC is a participant in the global level of world economy. Creating its own autonomous economic system, the legal superstructure of which is transnational law, TNCs become the main "operators" in international economic relations (Shumilov 2003).

9 The Practice of Resolving International Commercial Disputes Between TNCs in the Energy Sector: Categories and Incidents

The prevailing participation of TNCs as subjects of foreign trade in energy resources and their impact on its legal regulation proves the study of an array of international commercial disputes in the field of energy, a significant number of which are disputes between energy companies.

In the world energy sector there are four categories of disputes, depending on who acts as parties to the conflict:

1. State against the state. However, I must say that there is rarely a dispute between states in the energy sector. If they arise, then they are reduced to trade and territorial disputes (concerning boundary deposits). So, in 2013, the Philippines submitted to international arbitration a claim for a dispute with China over the territories in the South China Sea in accordance with the UN Convention on the Law of the Sea. In July 2016, the Permanent Court of Arbitration in The Hague found no grounds for China's territorial claims and considered that Beijing can not claim an exclusive economic zone in the Spratly archipelago area (RIA Novosti official website 2017).

   Trade disputes between states may arise with respect to measures that limit: (1) the transit of energy products (for example, monopolization of access); (2) import of energy products (for example, duties and taxes); (3) export of energy products (for example, export quotas), as well as in the provision of services in the energy sector;

2. The company is against the state. Disputes between TNCs and the state are primarily international investment disputes that arise as a result of significant changes in terms of the time of investment, expropriation or unjustified nationalization

   As an example, you can bring a dispute between Argentina and the Spanish oil company Repsol. In 2012, the Argentine authorities nationalized YPF (a subsidiary of Repsol). As a result, Repsol lost almost half of the proved reserves and 20% of the annual profit. In response, the company filed a suit with the International Arbitration Court demanding that Buenos Aires either return the assets or pay compensation in the amount of $ 10.5 billion. The dispute lasted almost two years. In 2014, the Board of Directors of Repsol accepted the proposal of the Government of Argentina for compensation of $ 5 billion for the nationalization of its assets (Vedomosti 2017). Under the agreement, approved by the board of directors of the company, Repsol will receive about half of this amount.

   Investment disputes can be based on contracts concluded between the investor and the host state (such as production sharing agreements, service contracts), and/or agreements on the promotion and mutual protection of investments. Russia has similar agreements with more than 60 states.

   In addition to investment disputes, one should also mention the growing attention to the issues of environmental protection in the world, which is reflected in disputes between energy companies and host states.

   The most famous example of recent years are suits to the British company BP in connection with the accident at Deepwater Horizon deepwater rig in the Gulf of Mexico in April 2010 and the ensuing largest oil spill in US history, which turned the accident into one of the largest man-made disasters by adversely affecting the ecological situation (Anisimov and Ryzhenkov 2015);

3. The company is against the company. International commercial disputes between energy companies are currently the most common. Among them, two areas of the substance of such disputes can be distinguished:

(1) in a situation where the interests of participants in joint ventures are affected or violated in contracts such as joint development agreements, lease agreements, sales agreements, confidentiality agreements, etc.;
(2) in a situation where there is a conflict of interest between operators of oil and gas projects and service companies.
The circumstances of such disputes are usually confidential. As an example, the disputes that arose in 2016 between Gazprom and the Danish company DONG Naturgas A/S, the Polish PGNiG, the Turkish BOTA Petroleum Pipeline Corporation, the Dutch Gas Terra B.V. and Shell Energy Europe, based in London, on the revision of natural gas prices (Made for minds 2017);
4. Individual/group of persons against the company. Claims of individuals/groups of persons against TNCs in the energy sector arise in the event of damage/harm to health. For example, claims of Nigerian fishermen and farmers to the British-Dutch company Shell at the High Court of London (March 2012) and to the Royal Court of the Netherlands (October 2012) related to the oil spill in the Ogoniland region in 2008 and pollution of ponds where the local population fished.

All these disputes to a greater or lesser extent concern the interests of Russia as one of the largest producers and exporters of energy, energy resources and Russian multinationals, which continue to integrate into the global energy sector (Analytical Center under the Government of the Russian Federation 2017).

10 The Authority of International Legal Entities on the Development of Legal Regulation of Foreign Trade Turnover of Energy Resources

World tendencies of liberalization of foreign trade turnover, internationalization of production and capital have involved international legal entities and integration associations in the international economy and politics.

Speaking about the participation and influence of TNC activities on the development of legal regulation of foreign trade turnover of energy resources in the context of the need for international integration in the energy sector, one can not help turning to the provisions of the Energy Strategy of Russia for the period until 2030, approved by the RF Government Decree No. 1715-r of 13.11.2009 (Order of the Government of the Russian Federation 2009). According to this program act, which determines the perspective directions of development of international energy cooperation, diplomatic support of the interests of Russian fuel and energy companies abroad, as well as a list of measures and mechanisms of the state energy policy includes: active participation in the international negotiation process on energy issues, ensuring a balance of interests of importers, exporters and transiters of energy resources in international treaties and the activities of international organizations; the development of cooperation in the field of energy with the countries of the Commonwealth of Independent States, the Eurasian Economic Community, Northeast Asia, the Shanghai Cooperation Organization, the European Union, and other international organizations and states; development of new forms of international (including technological) cooperation in the energy sector; Russia's active participation in international cooperation in the development of the energy sector of the future (hydrogen energy, thermonuclear energy, use of energy from sea tides, etc.); coordination of activities in the world oil and gas markets with the countries of the Organization of Petroleum Exporting Countries and the Forum of Gas Exporting Countries; assistance in the formation of a single European-Russian-Asian energy space; assistance in ensuring a favorable and non-discriminatory regime for the activities of domestic energy and service companies (as well as foreign companies with Russian participation) in world markets, including their access to foreign energy markets and final energy markets; assistance in attracting foreign investments on mutually beneficial terms, primarily in technically complex and risky projects; ensuring access of Russian energy companies to the use of resources of the world financial markets, advanced energy technologies", etc.

It is obvious that an important role in the development of legal regulation of foreign trade turnover of energy resources is assigned to various kinds of international legal entities, organizations and integration associations. A striking confirmation of this influence is that many changes introduced in the Civil Code of the Russian Federation take into account the current trends in the development of conflict regulation, which are reflected, in particular, in Roma Regulation I.

The changes introduced in Sect. 6 of the Civil Code reflect the world trends of the most modern legislation on private international law. The main changes concern, above all, expanding the scope of autonomy of the will of the participants in relations with the foreign element, reducing the role of formal requirements for drawing up various documents, expanding the independence of legal entities and their participants in the choice of applicable law, etc.

The general thrust of the changes introduced demonstrates the creation by the legislator of favorable conditions for the liberalization of foreign trade regulation, including in the energy sector, reducing the direct impact of the state on foreign trade and leaving an important role as a regulator of relations for participants in international economic activity—the parties to the treaty.

A promising trend in international treaty practice is the inclusion in the treaties of an increasing number of discretionary norms, based on the principle of freedom of contract and the autonomy of the will of participants in legal relations.

The provisions of the conflict rules of the Civil Code of the Russian Federation also confirm the general trend of expanding the disposability of contractual obligations.

The newest codifications in the field of private international law and the practice of its application abandon the requirement that the right chosen by the parties should be related to the legal relationship, allow the parties to choose as the applicable law not only national law, but also transnational law (lex mercatoria), general principles of law and even international law. In the practice of concluding international commercial contracts, a phenomenon arose, such as the "internationalization" of contracts, which aims to remove the contract from the effect of national law and conclude contracts in the form of self-sufficient autonomous legal systems. Following the psychological theory of law, it can be argued that, based on the principle of autonomy of the will of a party to a particular contract, they can create a special system of intuitive law, the specifics of which should be taken into account when implementing and interpreting this contract (Merezhko 2012).

At the present time, acts of private legal unification, that is, recommendations issued by international governmental and non-governmental organizations and generalizing international commercial practices and customs (standard contracts, manuals on drawing up contracts, unified rules, model laws, etc.), which are addressed "directly by participants of international commercial turnover in daily contractual practice, as well as by the state when drafting their own legislation" (Vilkova 2002).

Thus, TNCs, on the one hand, are the main subjects of relations in the sphere of foreign trade turnover of energy resources, and on the other hand, they represent a powerful mechanism for influencing them. Actively influencing relations in the sphere of foreign trade turnover of energy resources, TNCs, through the conclusion of treaties on the basis of the principle of autonomy of will, form new relations, modify their existing forms, thereby creating a special self-sufficient autonomous legal system.

The decisive importance of the authority of international legal entities and associations on the development of national legal regulation of the foreign trade turnover of energy resources is connected with the interest of the Russian state in expanding the export of all major types of energy resources, products of their processing, as well as high technologies in which Russian energy and industrial companies have competitive advantages.

11   Conclusion

In conclusion, it should be noted that the prevailing participation as the main subjects of the foreign trade turnover of the energy resources of Russian TNCs is determined by their inherent characteristics, which give advantages over other subjects of foreign economic activity and allow it to be most effectively implemented in the sphere of international trade in energy resources. Among such signs are: economic integrity by a complex cross-border interconnected ownership relationship or other structure relationships; implementation of coherent activities at the national and international levels; legal multiplicity, manifested in the existence of a different legal form of independent and dependent structural units of a legal entity; various forms of ownership of a multinational enterprise; centralized management and control of the parent enterprise, ensuring consistency of actions.

Confirmation of the participation of TNCs in the foreign trade turnover of energy resources as the main subject of economic relations, which has a significant impact on its legal regulation, are the results of an analysis of their participation in foreign trade contract relations in the energy sector. Such a conclusion is made possible by the legal documents examined, officially published statistical data of the largest Russian multinationals, as well as materials of international jurisprudence reflecting the categories of disputes in the sphere of international circulation of energy resources, the subject and the parties to the judicial procedures examined.

In addition, the growing role of TNCs in the sphere of international turnover of energy resources is both a consequence and the reason for the obvious tendency to unite the interests of states on an economically subjective level, which leads to the emergence of integration processes in the field of legal regulation of the foreign trade relations arising between them, the orientation towards its liberalization and unification.

In connection with the conclusion made, it is suggested that in the process of improving and unifying the legal regulation of the activities of subjects of foreign trade contractual relations in the energy sector, on the emerging international trade (contractual) practice of TNCs, taking into account the signs and features of their legal nature and constituting an important part of modern international private-law energy regulation. Among them, the TNC's nationality, the international nature of its activities, the domination of the capital of a country, the unity of government, the unification of monopoly capital, the agreed policy, the implementation of a common strategy, economic unity and legal multiplicity of independent companies independent in the structure, the possibility of implementing a vertically integrated business, non-interference by TNCs in the internal affairs of the receiving state and intergovernmental relations, respect for the sovereignty of states over their natural resources, material values and economic activities; the right of the state to regulate and control the activities of TNCs, to participate in the distribution of profits and the nationalization of foreign ownership on its territory.

Integration processes of development and liberalization of foreign trade energy regulation are manifested not only through the formation of international contractual (commercial) practice (international commercial law) in the activities of TNCs among international legal entities. Various international organizations and integration associations take an important part in the legal development and unification of the foreign trade turnover of energy resources. This participation is caused by the need to develop new forms of international energy cooperation in foreign economic activity.

Summarizing the information obtained during the analysis of the activities of international legal entities, primarily the largest Russian companies—suppliers of energy resources and the tendencies of development of legal regulation of foreign trade turnover of energy resources formed under the influence of this activity, it can be concluded that the main tasks in the course of such development are:

- the involving representatives of the largest energy corporations in working groups created in the process of lawmaking and improving the legal support for foreign trade in energy resources in order to maintain a balance of interests of importers, exporters and transiters of energy resources;

- the development of international cooperation and cooperation within the framework of international integration associations with the participation of the Russian Federation in the field of energy with the aim of developing a unified substantive basis for regulating the relations of the participating countries related to the circulation of energy resources;
- the development of a new form of international cooperation and cooperation of special competence within the framework of priority for Russia intergovernmental integration associations in the field of energy with the purpose of developing recommendatory contractual bases for regulating the relations of the participating countries related to the circulation of energy resources.[1]

Thus, today the process of development and unification of the legal regulation of the foreign trade turnover of energy resources can not ignore the impact of the activities of the main subjects of international economic activity in the field of energy—international legal entities. Based on the analysis of the characteristic features of TNCs, international jurisprudence and the established fact of their predominance as subjects of the foreign trade turnover of energy resources, it can be asserted that through the activities of TNCs integration processes of development and liberalization of energy regulation, as a result of which the role and the foundations of international private law treaty practice—the international commercial law of legal entities—energy companies, which is gaining increasing importance in the general legal array and should be taken into account in the process of lawmaking.

Under the influence of the activities of international legal entities and integration associations, the main prospective trends in the development of legal regulation of the foreign trade turnover of energy resources in the Russian Federation are being formed, evidence of its liberalization, a reduction in the direct impact of the state, an increase in the role of recommendatory acts of private legal unification, developed by international organizations and generalizing international commercial practices and customs, increasing the role as a regulator of relations between participants in international economic activity—the parties to the contract. Among the promising trends in the development of conflict regulation of foreign trade turnover of energy resources: the expansion of the scope of autonomy of the will of participants in relations with a foreign element, the reduction of the role of formal requirements for the compilation of various documents, the expansion of the independence of legal entities and their participants in the choice of the applicable law, the expansion of the disposability of contractual obligations, a departure from the requirement that the right chosen by the parties should relate to the legal relationship, the possibility

---

[1]To date, the International Energy Agency (IEA) is an autonomous international body within the Organization for Economic Cooperation and Development (OECD). The organization was established in Paris in 1974 and has 29 member countries. The main objective of the organization is to promote international cooperation in improving the global structure of supply and demand for energy resources and energy services. IEA advocates the interests of energy importing countries, and only OECD member states can become members of the IEA (Note by the authors).

of the parties choosing as the applicable law not only national law, but also the rights of international commercial contracts (lex mercatoria), as well as the general principles of law and international law.

A particularly significant contribution to the unification of the interests of states at the level of economic entities, the development and creation of a uniform private-law regulation of foreign trade relations in the energy sector in the context of expanding international cooperation, Russia is making international integration associations with the participation of the Russian Federation and/or its main foreign economic partners. In the light of this thesis, the following chapters of this study are devoted to the study of the legal regulation of the foreign trade turnover of energy resources in the international integration associations of the EU and BRICS.

References

Analytical Center under the Government of the Russian Federation. (2017). *Energy Bulletin*, (14). Retrieved March 5, 2017, from http://ac.gov.ru/files/publication/a/2992.pdf.

Anisimov, A. P., & Ryzhenkov, A. Ja. (2015). Linking environmental legislative inefficiency to lack of an environmental philosophy. *Environmental Policy and Law*, (3–4), 145–155.

Asoskov, A. V. (2000). Problems of legal regulation of transnational companies. *Juridical World, 8,* 43–47.

Boguslavsky, M. M. (1999). International Private Law. Moscow: Publishing house "Lawyer".

Boguslavsky, M. M. (2010). International Private Law. Moscow: Publishing house "Justicinform".

Calderon, H. L. V. (1984). Code of Conduct for Transnational Corporations and International Law (exemplified by the activities of transnational corporations in Latin America) dis. Cand. Jur. Sciences, Moscow.

Chuku, E. (2013). Transnational corporations in international energy relationships. *Power, 4,* 115–117.

Decree of the Government of the Russian Federation. (2014). On approval of the state program of the Russian Federation. Energy Efficiency and Energy Development, (approved by the Government of the Russian Federation No. 321).

Decree of the President of the Russian Federation. (2016). On approving the foreign policy concept of the russian federation, (approved by the Decree of the President of the Russian Federation on November 30, 2016, No. 640).

Doronina, N. G., & Semilutina, N. G. (2011). International private law and investment: scientific-practical research. Moscow: Kontrakt Publishing House.

Doronina, N. G., & Semilutina, N. G. (2012) Private international law and investments. Moscow: Publishing House Institute of Legislation and Comparative Law at the Government of the Russian Federation.

Federal Law. (1998). On investment activity in the Russian Federation, implemented in the form of capital investments, (adopted by the State Duma on July 15, 1998, No. 39-FZ).

Federal Law. (2003). On the fundamentals of state regulation of foreign trade activity, (adopted by the State Duma on November 21, 2003, No. 164-FZ).

Federal Law. (2005).On Concession Agreements, (adopted by the State Duma on July 6, 2005, No. 115-FZ).

Federal Law. (2007). On the State Atomic Energy Corporation "Rosatom", (adopted by the State Duma on November 13, 2007, No. 317-FZ).

Fedchuk, V. D. (2010). *De facto dependence of de jure independent legal entities: Penetration of the corporate curtain in the law of leading foreign countries*. Moscow: Volters Kluver.

Gazprom. (2017a). About Gazprom. Retrieved January 10, 2017, from http://www.gazprom.ru/about/marketing/europe/.

Gazprom. (2017b). Alexey Miller informed Vladimir Putin about the results of Gazprom's work in 2016. Retrieved January 10, 2017, from http://www.gazprom.ru/press/news/2017/january/article299819/.

Gazprom in questions and answers. (2017). Gazprom on foreign markets. Retrieved March 10, 2017, from http://www.gazpromquestions.ru/foreign-markets.

Gazprom International. (2017). Partners. Retrieved March 5, 2017, from http://gazprom-international.com/ru/operations/partnery.

Inshakova, A. O. (2008). Unification of corporate regulation in the European Union and the Commonwealth of Independent States dis. Doct. Jurid. Sciences, Moscow.

Inshakova, A. O. (2012). Peculiarities of defining the concept and legal personality of transnational corporations in international and national lawmaking and doctrine. *International Public and Private Law, 4,* 10–13.

Inshakov, O. V., Kleiner, G. B., & Sorokozherdiev, V. V. (2014). Strategy for modernization of the Russian Economy: Theory, policy, practice of implementation. Moscow: Publishing House "Modern Economics and Law".

Kamalov, M. M. (2011). Ineffectiveness of the convention of the commonwealth of independent states "On Transnational Corporations". *International Public and Private Law, 2,* 10–13.

Kozlov, S. V. (2014). Legal regulation of domestic trade in oil and oil products markets. *Energy Law, 2,* 38–41.

Krivorotko, I. A. (2015). Problems and prospects of Russian Gas TNCs on the world market, innovative. *Science, 9,* 171–176.

Kulagin, M. I. (1997). State-monopoly capitalism and a legal person, Statute, Moscow.

Kulikov, R. A. (2004). On the definition of criteria for the concept of TNCs. *International Public and Private Law, 2*(17), 3–11.

LUKOIL. (2017). Annual Report of PJSC "LUKOIL" for 2015. Retrieved January 10, 2017, from http://www.lukoil.ru/Business/Upstream/KeyProjects.

Lunts, L. A. (1976). Multinational enterprises of capitalist countries in the aspect of private international law. *Contemporary State and Law, 5,* 123–124.

Lysenko, D. L. (2003). Problems of the legal status of transnational corporations: International legal aspects: The author's abstract. Cand. Jurid. Sciences, Moscow.

Made for minds. (2017). Why does half of Europe argue in court with Gazprom because of gas prices. Retrieved March 33, 2017, from http://www.dw.com/ru/WhydoeshalfofEuropearguein courtwithGazprombecauseofgasprices/a-19265700.

Merezhko, A. A. (2012). The psychological theory of private international law. *Russian Journal of Law, 6,* 52–68.

Natalukha, V. V. (1985). International Private Business and the State, International Relations, Moscow.

Official website of the Ministry of Energy of the Russian Federation. (2017). Approved the forecast of scientific and technological development of the fuel and energy sector of Russia for the period until 2035. Retrieved January 26, 2017, from http://minenergo.gov.ru/node/6365.

Osminin, B. I. (1983). Transnational corporations and international law: An abstract of dissertations. Cand. Jurid. Sciences, Moscow.

Official website RIA Novosti. (2017). South China Sea dispute: The court in the hague rejected the rights of Beijing. Retrieved March 5, 2017, from https://ria.ru/world/20160712/1464640532.html.

Official site Rosneft (2017). Rosneft today. Retrieved January 10, 2017, from https://www.rosneft.ru/about/Glance/.

Order of the Government of the Russian Federation. (2008). On the Concept of Long-Term Social and Economic Development of the Russian Federation for the Period to 2020, (approved by the Government of the Russian Federation on November 17, 2008, No. 1662-r).

Order of the Government of the Russian Federation. (2009). On the Energy Strategy of Russia for the period up to 2030, (approved by the Government of the Russian Federation on November 13, 2009, No. 1715-p).

Parfenov, I. A. (1999). Management of holding in oil and gas complex (legal aspects): dis. Cand. Jurid. Sciences, Tyumen.

Romanova, V. V. (2014). *Energy Law: A common part. Special part: Textbook.* Moscow: Publishing house "Lawyer".

Rosatom (2017). About Rosatom. Retrieved March 4, 2017, from http://www.rosatom.ru/about/.

Schmitthoff, C. M. (1954). *The English Conflict of Laws* (3rd ed.) London.

Schmitthoff, C. M. (1973). The multinational enterprise in the United Kingdom, Nationalism and the multinational enterprise: legal, economic and managerial aspects (p. 24).

Scholarship Repository (2017). Weissbrodt, D. & Kruger, M. Norms on the Responsibilities of Transnational Corporations and Other Business Enterprises with Regard to Human Rights. Retrieved March 5, 2017, from https://scholarship.law.umn.edu/cgi/viewcontent.cgi?article=1247&context=faculty_articles.

Sharkova, E. A., Gavra, D. P., Shishkin, D. P., Pankova, G. K., & Glinternik, E. M. (2016). Transnational companies and local communities: Aspects informational interaction. *Man in India, 96*(10), 3805–3815.

Shevtsova, O. I. (2008). Legal nature of transnational corporations. *Journal of International Law and International Relations, 4,* 7–15.

Shiyanov, A. V. (2016). Russian transnational corporations in the system of international legal regulation of regional energy markets. *Eurasian Juridical Journal, 1,* 137–140.

Shumilov, V. M. (2003). *International economic law in the age of globalization.* Moscow: International Relations.

The Eurasian Economic Commission. (2017). The Treaty on the Eurasian Economic Union. Retrieved May 5, 2017, from http://www.eurasiancommission.org/.

Tripartite Declaration of Principles Concerning Multinational Corporations and Social Policy. (2001). International Labor Office Publishers, Geneva.

Vedomosti (2017). Argentina will pay Repsol $ 5 billion for the nationalization of its assets. Retrieved March 5, 2017, from http://www.vedomosti.ru/business/articles/2014/02/26/ArgentinawillpayRepsol$5billionforthenationalizationofitsassets.

Velyaminov, G. M. (2015). International Law: Experiences, Statute, Moscow.

Vilkova, N. G. (2002). *Contract law in international circulation.* Moscow: Statute.

Woidasky, J., & Hirth, T. (2012). Resource efficiency from today to beyond tomorrow. *Chemie-Ingenieur-Technik, 84*(7), 969–976.

World Investment Report. (2001). *Promoting linkages.* Geneva: UNCTAD.

World Investment Report Transnational Corporations and Export Competitiveness. (2002). UNCTAD, Geneva.

Participation of International Organizations and Integration Associations in the Development of Legal Regulation of Foreign Trade in Energy Resources

Agnessa O. Inshakova, Elena I. Inshakova and Igor P. Marchukov

1 Annotation

The chapter is devoted to the study of the role of international organizations and integration associations, as a special category of participants and subjects of law-making in the sphere under study.

It is noted that international organizations and integration associations are actively involved in the development, improvement and unification of civil-law regulation of foreign trade in energy resources due to the impossibility of observing the world balance of interests of economic entities of producing countries, suppliers, consumers and transit countries of energy resources only through domestic methods. The role of such entities in the formation of legal regulation in the field under investigation is constantly increasing, this is due to the acquisition by them of ever greater legal personality and the need for operational lawmaking in the sphere of foreign trade in energy resources.

It is asserted that acts of international organizations and integration associations accumulate experience of states in the field of energy, contain specific provisions and new approaches, in comparison with the provisions of existing international

A. O. Inshakova (✉)
Department of Civil and International Private Law, Institute of Law,
Volgograd State University, Volgograd, Russia
e-mail: gimchp@volsu.ru; ainshakova@list.ru

E. I. Inshakova
Department of Economic Theory, World and Regional Economics,
Institute of Economics and Finance, Volgograd State University, Volgograd, Russia
e-mail: inshakovaei@volsu.ru

I. P. Marchukov
Department of Civil and International Private Law, Volgograd State University,
Volgograd, Russia
e-mail: gimchp@volsu.ru

© Springer International Publishing AG, part of Springer Nature 2019
O. V. Inshakov et al. (eds.), *Energy Sector: A Systemic Analysis of Economy,
Foreign Trade and Legal Regulations*, Lecture Notes in Networks and Systems 44,
https://doi.org/10.1007/978-3-319-90966-0_12

legal norms, push states in their work on codification and improvement of the civil-law sphere.

It is substantiated that the sectoral bodies of international organizations and integration associations have great opportunities to form an integration and competitive model of the state energy policy, which should be recognized as a priority in modern conditions, and one of the main tasks of which is the creation of unified legal regulation. In this regard, the issues of the unification of civil and legal regulation of the foreign trade turnover of energy resources are further considered in the work on the example of EU countries, as a vivid example of the formation of an integrationally competitive model for the development of energy regulation within the framework of the international integration association of Europe, the main foreign economic partner of the Russian Federation.

The study argues the advantages of international integration associations before other international organizations and entities involved in the process of unifying the foreign trade turnover of energy resources, which are due to a higher degree of interest, common interests and harmony achieved in the relationship, as well as a more developed and cohesive institutional and organizational structure.

2 Materials

Opinion on the international institutes and branch bodies of major international entities operating in the energy sector was formed on the basis of international legal acts and official data of regional integration associations (EC, CIS), international organizations of universal competence (UN, UNESCO), universal international organizations of special competence (WTO, OPEC, IEA), regional international organizations of special competence (OAPEC, OLADE), non-governmental organizations (MIRES).

The role, targets, methods of influence and activities of various international organizations and integration associations involved in the development, improvement and unification of legal regulation of foreign trade turnover of energy resources were studied using the example of the works of Anufrieva et al. (2009), Kuzmin and Kagramanov (2009), Shcherbakova (1999), Panova (2015), Romanova (2013).

The problems of liberalization of the energy industry, deregulation and efficiency of modern state energy policy models have been investigated through scientific work (Bazhanov 2003).

The creation of the author's position on the significance of the activities of international organizations and their bodies in the energy sector, including those related to foreign trade in energy resources, was provided by Grishaeva (2011), Gudkov (2016), Kukushkina (2014), Nartova and Matteotti-Berkutova (2012), Novikova (2006), Ruzhin (2013).

The work of Kovaleva (1999) contributed to the study of the law-making activity of interstate organizations.

Judicial practice of the Russian Federation as an energy exporting country in relations with the EU was studied on the example of cases considered by the WTO Dispute Resolution Authority: No. T-459/08 "EuroChem" against the EU Council, the European Commission and "Fertilizers Europe"; No. T-84/07 "EuroChem" against the EU Council, the European Commission and "Fertilizers Europe"; No. T-235/08 of Acron and Dorogobush against the EU Council, the European Commission and Fertilizers Europe; No. T-118/10 "Acron" against the EU Council, the European Commission and "Fertilizers Europe".

3 Methods

The scientific research carried out in this part of the monograph was based on the methodological basis represented by the general scientific method of historical materialism. In addition, general scientific methods of cognition were used in the work: analysis, generalization, classification, induction, deduction, synthesis, dialectics, statistics, SWOT analysis, etc. Among the particular scientific methods used in the study, a special place is given to comparative legal analysis and the method evaluation of legal processes.

4 Introduction

The need for supervision by various kinds of international organizations and associations of lawmaking and unification of foreign economic energy regulation is now evident for states that are actively involved in the integration processes and realize the impossibility of further preserving the world balance of interests of economic entities of producing countries, suppliers, consumers and transiters of energy resources only through domestic methods.

At the end of the twentieth century in the system of international economic relations, the role of interstate interaction in energy regulation has significantly increased. The energy crisis of 1997–2000, which led to a sharp price fluctuation in the oil markets and a serious upheaval in the world energy sector, ultimately led to the realization of the need for a significant increase in international energy policy. Modern energy diplomacy includes a whole range of issues, including reliable energy supply to consumers, access to their sources, transportation routes for hydrocarbon raw materials and electricity, and international aspects of nuclear energy. Today this is a significant direction of foreign policy activity, contributing to the creation of conditions for normal interstate relations in the energy sector. In addition, it is necessary to take into account the growing influence on the energy diplomacy of the negative consequences of globalization, including global finance, ecology, etc.

We believe that the crisis in the energy sector as a result of reforms to liberalize the energy industry, rising prices and emergency power outages that shocked in the early 2000s. England, Canada, Argentina, Brazil, Uruguay and Colombia (Anufrieva et al. 2009) speak of the short-term nature of the positive results of deregulation and liberalization in the gas and electricity sectors. Therefore, there is not a cessation, but an increase in crisis phenomena. This is confirmed by the fact that competitive gas markets, as well as electricity markets, failed to develop mechanisms for attracting investments to finance large-scale energy projects and were faced with the strongest deficit of energy resources and, as a result, with price increases (Bazhanov 2003).

5 Improving Energy Efficiency and a Modern Model of State Energy Policy

The example of many countries shows that the introduction of the so-called national-competitive model of the state energy policy (Anufrieva et al. 2009) through the deregulation of previously considered monopolistic markets, primarily gas and electricity, as well as the liberalization of energy industries is not able to solve the main problem—involvement of private investment in the industry and increase of efficiency and reliability of energy supply to consumers.[1] And the main arguments of the reformers—the reduction of energy prices and the increase in the efficiency of the energy sector—do not stand up to any criticism: prices are rising, and the promises of greater efficiency lie in the banal profits of private companies.

Unable to cope with the crisis situation, the national-competitive national-monopoly model of energy policy, which led to a dead end energy markets through the unacceptable desire in the globalization of the economy for a directive model of energy management, as well as state control of the main resources by preserving the leading assets in the energy sector. Within this model, energy policy is formed in isolation from the economy as a whole.

We believe that such an option for solving energy problems should be sought, which is most consistent with the confirmed trends in the world economic and legal development under globalization conditions (Baldwin et al. 2017). According to this option, the solution of the problems of the energy sector lies in the integration of state energy policies, as well as the development of common standards and governance laws in this area. Today, scientists are talking about the formation of a new model of

---

[1]In the scientific doctrine, three models of state energy policy are singled out. The first model—national monopoly—is typical for states that have significant energy resources. Its distinctive feature is the attitude to energy policy as a separate economic segment, largely controlled by the state. The second model is national-competitive, where the dominant position is taken by the liberal doctrine, which calls for a reduction in state intervention in all sectors of the economy. And, finally, the new, emerging third model of the state energy policy is integration-competitive, designed to create a common energy market and to develop a unified state policy for a number of countries (*Note of the authors*).

state energy policy—integration-competitive, designed to create a common energy market and develop a unified state policy for a number of countries.

The integration—competitive model is most graphically represented in the energy policy of the European Union and Scandinavian countries (Norway, Sweden, Denmark, Finland). The state energy policy in these states is based on the continuation of the general policy of development of competition and integration of the European market (Kuzmin and Kagramanov 2009). In this connection, the issues of unification of the legal regulation of the foreign trade turnover of energy resources in this work will be examined, for example, in the EU countries as a vivid example of the formation of an integration and competitive model—most promising from the point of view of the authors and developing within the framework of the international integration association of Europe, the main foreign economic partner of the Russian Federation in this field (Hooimeijer and Tummers 2017).

An important condition for the implementation of a unified energy strategy and the formation of a single energy market continues to be the liberalization of energy markets and the development of competitive relations in the energy sector. It should be noted that the nationally competitive and integration-competitive models of the state energy policy are quite similar in their basic characteristics. The latter is the original outcome of the evolution of the national-competitive model. The characteristic features of both models is the development of competitive relations in the energy sphere (Shcherbakova 1999). The role of the state in these models is quite high, but the emphasis is on creating conditions for the functioning of the market (in particular, on creating a regulatory and legal framework), and not on using administrative resources and directing management. At the same time, the specificity of the integration-competitive model is to reduce the role of national state policies in the energy sector.

6 International Organizations and Integration Associations Operating in the Energy Sector: Role, Goals, Classification

The effectiveness of functioning of integrated energy institutions has a huge impact on the state of the entire system of modern international economic relations. In this regard, the states, acting as subjects of the legal regulation of the energy sector, unite efforts and create international organizations, as well as various other forms of interstate interaction with the aim of effectively managing the world fuel and energy complex and maintaining the "energy balance". The role of such entities in international law is constantly growing, this is due to the acquisition by them of ever greater legal personality. The increase in the volume of legal personality of international associations is inevitable, necessary for the realization of their goals, including, in matters of convergence of energy regulation of various nationalities. To such purposes it is necessary to carry:

- the formation of a common agreed position of the participating countries on the development of international energy relations;

- development of common rules for the behavior of participants in energy legal relations, as well as projects, program acts, agreements, standards, technical regulations and other documents aimed at standardization and bringing to uniform legal regulation;
- harmonization of international economic regulation in the energy sector.

The activities of international organizations and integration associations are often more effective than the activities of individual countries. According to the researchers, this fact is due to the involvement of qualified representatives of various states, independent experts and industry specialists in the decision-making process, which allows solving the problems of the fuel and energy complex more quickly and effectively (Romanova 2013). The influence on a wide range of energy relations through the activities of international organizations and associations is also helped by the fact that the states that create them have a special interest in shaping the common legal space and are active participants in the foreign trade turnover of energy resources.

At present, there are a huge number of different international organizations influencing the formation of the world energy policy. But, not all specialists express pleasure in this matter. So, N. Dubash and A. Florini point out that "the world energy landscape is littered with governing institutions" (Panova 2015).

In the scientific literature, researchers seek to differentiate such organizations and associations into specific groups for various reasons. There are various approaches to the classification of international entities operating in the energy sector and the criteria that are the basis for their differentiation. So, the basis of classification can be the nature of the influence of the organization or association on international energy relations, which can be direct or indirect. An international organization with direct influence is OPEC.[2] This organization carries out foreign economic legal regulation of the energy industry through direct impact on the world fuel market. The indirect nature of the impact on international energy relations lies in the regulation of areas of activity adjacent to the energy sector (Kukushkina 2014). An example is the work of UN apparatus, the CIS or UNESCO in the field of ecology, since energy is one of its main pollutants, and therefore international environmental standards will have significant consequences for the work of this sector.

Another classification is based on the subject composition. According to this approach, there are governmental and non-governmental international organizations in the energy sector. Members of government organizations are states represented by authorized representatives. They constitute the majority of international entities. Non-governmental organizations are structures formed by private groups of individuals or individuals of a number of countries with specific objectives. It is important to distinguish such organizations from another type of international legal entity—transnational corporations, whose purpose is to make a profit. In the energy sector, non-governmental organizations play a significant role, as the World Energy Council (WEC) conducts regular conferences on pressing energy issues, organizes

---

[2]The organization of oil-exporting countries is an international intergovernmental organization set up by oil-producing countries in order to control oil production quotas (*Note by the authors*).

interaction among participants in the energy market, and promotes its development. As a rule, the main members of such organizations are large energy companies, primarily in the gas and oil industries.

However, the most rational, in our opinion, is the classification by the specifics of the activities of international organizations and associations in the energy sector. On the basis of this criterion, the following categories of such formations can be distinguished:

- universal international organizations of special competence (OPEC, IEA[3]);
- regional international organizations of special competence (OAPEC,[4] OLADE[5]).

7 The Influence of International Entities and Their Branch Bodies on the Process of Unification of Energy Law

The active activity of international organizations and associations in the field of energy contributes to the development of unified approaches to the regulation of private relations.

Firstly, international organizations and integration associations contribute to the rapprochement of the positions of the member countries of such associations, stipulating the development of agreed international legal documents.

Secondly, they themselves take an active part in international lawmaking through the implementation of their own norm-setting activities. This activity is aimed at the development and adoption of international treaties as a party and is also manifested in the process of "independent lawmaking", implemented through the adoption of resolutions containing international legal norms (Kovaleva 1999).

Thirdly, international organizations and associations carry out various organizational, executive, supervisory and judicial functions in order to promptly and fully implement international legal regulations, as well as to bring to justice those responsible for their violation of subjects (Gavrilov 2004).

International organizations and integration associations help to identify global problems in the energy sector and find ways to address them through the adoption of mutually agreed acts.

---

[3]An autonomous international apparat within the framework of the Organization for Economic Cooperation and Development (OECD), established to interact in improving the world structure of energy supply and demand and energy services (*Note by the authors*).

[4]The Organization of Arab Oil Exporting Countries is a cartel created by the oil-producing countries to stabilize oil prices (*Note by the authors*).

[5]Latin American Energy Organization—Latin American Inter-State Organization for the Coordination of Energy Development (*Note by the authors*).

In the largest universal international organizations and integration associations (EU, CIS, BRICS, EEU) there are or are planned to create branch apparatus.

Thus, within the framework of the CIS, an international sectoral energy organization has been set up—the Electric Power Council, which consists of the heads of relevant state authorities and national electric power companies of the participating states, which are given appropriate powers by states.

The CIS Electric Power Council was established by the Intergovernmental Agreement on the Coordination of Interstate Relations in the Electric Power Industry of the CIS of February 14, 1992.

The permanent working apparat of the CIS Electric Power Council is the Executive Committee, located in Moscow.

The main tasks and functions of the Council include, among others:

- development of proposals on the legal conditions for ensuring the joint operation of the unification of the electric power systems of the CIS member states;
- assistance to the CIS member states in the unification and harmonization of regulatory legal acts in the electric power industry;
- development of proposals on the directions of integration of the CIS member states in the field of electric power industry;
- participation in the preparation of international agreements in the energy sphere;
- development of international relations in the interests of the CIS member states and participation in the work of international energy organizations and others (CIS Electric Power Council 2017).

The CIS Electric Power Council (hereinafter—EES) is engaged in law-making activities. So, within the framework of his activity, "Unified principles of parallel operation of energy systems" were defined, which became one of the foundations of the "Agreement on the parallel operation of CIS energy systems", as well as the "Concept (Basic Principles) for the Construction and Operation of the System for the Intergovernmental Exchange of Scientific and Technical Information in the Field of the Electric Power Industry of the CIS". The CIS EES developed the terminology for the Agreement on the Transit of Electricity and Power of the CIS Member States, the Agreement on Mutual Assistance in Cases of Accidents and Other Emergencies, and the Treaty on the Provision of Parallel Operation of Electric Power Systems of the CIS Countries. The CIS EES created the "Provisional Regulation on the Procedure for Calculating Tariffs for Transit", developed the "Concept for the Formation of the Common Electric Power Market of the CIS Member States". The EES of the Commonwealth of Independent States (CIS) exercises overall leadership in the creation, functioning and development of the Intergovernmental System for the Exchange of Scientific and Technical Information in the Field of Electricity (Novikova 2006).

As another example, let us turn to the integration of the EU. In the structure of the European Commission there is a special department responsible for the issues of European energy, the Directorate General for Energy. The Commission also established a number of specialized agencies in this field, such as the European

Energy Forum (2017) or the European Energy Regulators (Council of European Energy Regulators 2017).

The Directorate-General for Energy is responsible for the development and implementation of the EU energy policy and performs a number of tasks: monitors the energy market; Strengthens the energy infrastructure; Promotes innovation in the field of energy technologies; develops the most advanced legal framework for the use of nuclear energy, etc. In all areas, the Directorate develops short-term, medium-term and long-term energy policies; monitors the implementation of current EU legislation; encourages the exchange of best practices; provides information to interested parties (Official website of the EU 2017).

The General Directorate of Energy consists of five departments, consisting of 17 separate units.

At the EU summit held in March 2015 in Brussels, the initiative to create the EU Energy Union was unanimously approved. The new organization is called upon to become a single European supranational institution regulating EU energy relations, both with third countries and with non-European energy companies (Foreign Policy 2017).

Under the auspices of the Energy Union, the Commission actively promotes initiatives to strengthen supranational powers in the fuel and energy complex, both in the regulation of the internal energy market and in the field of external energy relations (Gudkov 2016).

In the framework of the BRICS unification, there is currently no sectoral body in the sphere under consideration. At the same time, in order to interact with other BRICS member states in the energy sector, including the development of the legal framework for international energy cooperation (the Concept of the Participation of the Russian Federation in the BRICS Union 2013), President of the Russian Federation V.V. Putin in the summer of 2014 proposed the creation of the BRICS Energy Association, whose activities will be aimed at ensuring the energy security of the participating countries.

The creation of such sectoral bodies within the framework of international organizations and integration associations, the strengthening of their activity at the international level, speaks about the increasing political importance of energy issues, the recognition by states of the role of international organizations and associations in the development of international cooperation and the unification of legal regulation.

8 Formation of a Unified Energy Policy and Methods for the Unification of Foreign Trade Energy Regulation

In the modern period, the actual task of international organizations and integration associations is the convergence of the legal principles of regulation in the field of energy. This is facilitated by the implementation of joint projects in the energy sector, strategies, plans, recommendations, joint scientific developments aimed at harmonizing the legal regulation.

Acts of international organizations and integration associations accumulate the experience of states in the field of energy, contain specific provisions and new approaches in comparison with the provisions of existing international legal norms, push states in their work on codification and improvement of international law, including in the civil law sphere.

Considering all the above, it should be concluded that in the modern period the importance of normative and recommendatory acts of both universal and regional international organizations and integration associations in the regulation of international relations in the energy sector is steadily increasing. The main reason for this is the need for operational regulation of relations in the energy sector, which can be achieved, including by improving the process of lawmaking of international organizations and integration associations.

Successful integration is impossible without an appropriate global legal framework consisting of three levels: national legal level, which should be harmonized with legal acts of regional integration associations, forming the second level of legal regulation; the third level of legal regulation presupposes achieving legal harmonization between the regional level and the global regulation of economic and trade relations. Such a global level in the sphere of foreign trade turnover, including energy resources, is the right of the WTO.

After Russia joined the World Trade Organization (WTO) in August 2012, it assumed certain obligations. Despite the activities of many specialized international organizations, such as OPEC, the Energy Charter Treaty, the International Energy Forum, we can state a regulatory and legal weakness with regard to the energy market with Russia's participation. Due to the fact that multilateral trade agreements of the WTO do not exclude trade in resources, their provisions related to access to markets, transit issues, energy services, ecology, trade barriers, subsidies, investments and technology exchange are applicable to the energy sphere. However, the agreements currently in force do not cover all aspects of energy trade.

As experts rightly point out, the specificity of energy trade is that export duties are coming to the fore here, which are practically not regulated by agreements within the WTO, since historically resource-rich countries have hindered the achievement of any agreements on export barriers (Kukushkina 2014). Consequently, the commodity turnover of energy carriers by the WTO system is not regulated. The situation with the regulatory system is especially complicated in the gas market. There are opinions that this is due primarily to the need to organize special conditions for trade in this sector, which are expressed in the creation of an appropriate infrastructure, addressing transit and investment issues (Grishaeva 2011). However, these issues are not regulated by the WTO agreement system.

The most important document regulating the issues of trade within the WTO is the General Agreement on Tariffs and Trade (GATT) (WTO Documentation Center 2017). Regarding the regulation of trade relations in the energy sector, GATT contains Article 5, which establishes the freedom of transit of energy resources. In particular, this is a ban on the delay or restriction of transit and the exclusion of various customs duties (Ruzhin 2013). Also, with the aim of harmonizing the external economic legal regulation of trade relations in the WTO, a specialized quasi-judicial Disputes Settlement Apparat (hereinafter—the DSA) was established. The purpose of the DSA is to resolve disputes between WTO members on issues that are not regulated by the multilateral trading agreement system.

So, with accession to the WTO, the Russian Federation filed a complaint against the EU's actions to recalculate the cost of energy products, as amended, in view of the special situation on the market of the exporting country. Anti-dumping duties apply to some Russian energy products on the EU market. That is, the Union does not take into account low gas prices, as a result of Russia's application of a double taxation system, which contradicts the provision of the Agreement of Partnership and Cooperation (APC) (EU portal 2017) on mandatory consideration of the factors of the competitiveness of the Russian economy. However, due to the fact, that the APC does not contain a mechanism of coercion and application of sanctions in this case, Russian energy companies PJSC Evrokhim and Acron forced to apply for restoration of the violated right to the WTO Disputes Settlement Body (Official Journal of the European Court of Justice 2017). Thus, the Disputes Settlement Apparat, acting as a quasi-judicial instance, is a kind of regulator in the field of trade in energy resources, and international trade in energy and energy services regulated by the WTO rules. Such goods and services fall under the basic principles of non-discrimination and market access provided for in the basic documents of the organization, such as: the General Agreement on Tariffs and Trade (GATT), the General Agreement on Trade in Services (GATS), the Agreement on Trade-Related Aspects of Intellectual Property Rights and the Agreement on Subsidies and Countervailing Measures. Energy is no exception, despite the fact that the WTO does not yet have a separate agreement that specifically would regulate international trade in gas, oil, renewable energy sources and other energy sources and energy services (Nartova and Matteotti-Berkutova 2012).

In this regard, the most important task of the World Trade Organization at this stage is the development and adoption of an agreement on sustainable energy within the framework of the existing GATT, which will allow for a course on harmonization of trade relations between WTO members.

The agreement should find a reflection of the situation:

- promoting the legal provision of energy as a special integrated market sector;
- on the classification of energy goods and services. At the same time, trade in all types of energy carriers should be regulated by the same norms;
- clarifying the rules of competition and public procurement;
- defining rules of subsidizing;
- setting a clear framework for restrictions in energy production and energy exports.

9 Conclusion

Taking into account the proposed classification on the basis of the specificity criteria of the activities involved in the formation of a unified energy policy and the unification of foreign trade energy legal regulation, it is established, that international organizations and integration associations are active participants in international lawmaking. Acts of international organizations and integration associations accumulate the experience of states in the field of energy, contain specific provisions and new approaches, in comparison with the provisions of the existing international legal norms, are pushing states in their work on the codification and improvement of international law, including in the civil law sphere.

It substantiated that the sectoral bodies of international organizations and integration associations have great opportunities to form an integration and competitive model of the state energy policy, which recognized as a priority in modern conditions and one of the main tasks of which is the creation of a unified energy legal regulation.

The creation of such sectoral bodies within the framework of international organizations and integration associations, the strengthening of their activity at the international level, speaks about the increasing political importance of energy issues, the recognition by states of the role of international organizations and associations in the development of international cooperation and the unification of legal regulation.

The advantages of international integration associations in front of other international entities involved in the process of unifying the foreign trade turnover of energy resources are due to a higher degree of interest, common interests and harmony achieved in the relationship, as well as a more developed and cohesive institutional and organizational structure.

References

Anufrieva, L. P., Bekiashev, C. A., Moiseev, E. G., & Ustinov, V. V. (2009). *International public law*. Moscow: Prospekt.

Baldwin, E., Rountree, V. et al. (2017). Distributed resources and distributed governance: stakeholder participation in demand side management governance. *Energy Research and Social Science*, (39), 37–45.

Bazhanov, E. P. (2003). *Actual problems of international relations*. Moscow: Science Book.

CIS Electric Power Council. (2017). Executive Committee of CIS EES. Retrieved May 8, 2017, from http://energo-cis.ru/rumain11/.

Concept of the participation of the Russian Federation in the BRICS Union (approved by the President of the Russian Federation on February 9, 2013). (2013). *Politics and Economics*, 1–2 (51), 39–43.

Council of European Energy Regulators. (2017). European Energy sector. Retrieved May 15, 2017, from http://www.ceer.eu/portal/page/portal/EER_HOME.

EU portal. (2017). Memorandum on industrial cooperation in the energy sector. Retrieved May 15, 2017, from http://www.russianmission.eu/userfiles/file/partnership_and_cooperation_agreement_1997_eng.pdf.

Foreign Policy. (2017). Formation of the EU Energy Union and the Interests of Russia. Retrieved May 16, 2017, from http://www.foreignpolicy.ru/analyses/obrazovanie-energeticheskogo-soyuza-es-i-interesy-rossii/.

Gavrilov, V. V. (2004). The concept of national and international legal systems. *Journal of Russian Law, 11,* 98–112.

Grishaeva, L. E. (2011). Russia and the WTO: modern braking mechanism. *Economic Journal, 23,* 6–8.

Gudkov, I. V. (2016). Competence of the European union for regulation of relations in the energy sector. *International Economic Law, 1,* 10–17.

Hooimeijer, F., & Tummers, L. (2017). Integrating subsurface management into spatial planning in the Netherlands. *Sweden and Flanders Proceedings of the Institution of Civil Engineers: Urban Design and Planning, 170*(4), 161–172.

Kovaleva, T.M. (1999). *Legislative activity of the interstate organization as a way to implement the founding act*, Kaliningrad.

Kukushkina, U.M. (2014). The importance of trade regulation within the WTO for the Energy Dialogue between Russia and the European Union. *Bulletin of the Financial University*, (4), 6–10.

Kuzmin, E.L., & Kagramanov, A.K. (2009). *Global energy security and pipeline transport, political and legal aspect*. Moscow: Scientific Book.

Nartova, O., & Matteotti-Berkutova, S. (2012). Russia's association to the WTO and energy. *Analytics and News on Trade and Sustainable Development, 3,* 18–20.

Novikova, S.N. (2006). Historical, political and legal features of the creation of the Electric Power Council of the countries of the Commonwealth of Independent States. *Law and Politics*, (5), 104–115.

Official Journal of the European Court of Justice. (2017). Case no. T-459/08 "EuroChem" against the EU Council, the European Commission and "Fertilizer Europe»; no. T-84/07 "EuroChem" against the EU Council, the European Commission and "Fertilizer Europe"; Case no. T-235/08 of Acron and Dorogobush against the EU Council, the European Commission and Fertilizer Europe; Case no. T-118/10 "Acron" against the EU Council, the European Commission and «Fertilizer Europe". Retrieved November 21, 2017 http://eur-lex.europa.eu/oj/direct-access.html.

Official website of the EU. (2017). Energy. Retrieved May 16, 2017, from http://ec.europa.eu/energy/en/about-us.

The official website of the European Energy Forum. (2017). Retrieved May 15, 2017, from http://www.europeanenergyforum.eu.

Panova, V. V. (2015). Global governance in the field of energy: Myth or reality? *Bulletin of International Organizations: Education, Science, New Economy, 10,* 33–39.

Romanova, B.B. (2013). Peculiarities of the legal nature in the field of energy. *The Legal World, 4,* 7–9.

Ruzhin, A.S. (2013). Regional trade agreements in the GATT/ WTO system: the role of the principles of international economic law. *Volgograd State University Bulletin. Series 5: Jurisprudence*, (1), 6–9.

Shcherbakova, M. A. (1999). *Interrelation of energy and economy*. Moscow: MEN.

WTO Documentation Center. (2017). General agreement on trade and tariffs of the WTO (GATT). Retrieved November 30, 2017, from http://base.garant.ru/2560614/.

The General Energy Policy and Ways of Development of Legal Regulation of the Foreign Trade Turnover of Energy Resources of the Russian Federation and the EU

Agnessa O. Inshakova, Evgenia E. Frolova and Igor P. Marchukov

1 Annotation

The chapter of the monograph is devoted to the analysis of the main directions of the development of the common energy law of the EU, the ways of their implementation through specific methods and mechanisms for unifying civil and legal regulation of the foreign trade turnover of energy resources.

The conflict rules of EU legal acts containing rules for determining the applicable law to obligations in the sphere of foreign trade turnover of energy resources investigated. Based on the analysis of key EU policy documents reflecting its energy policy, modern trends in the development of conflict-of-law regulation of contractual obligations revealed. It's substantiated that the EU does not depart from the course taken a few decades ago to gradually unify the legal regulation in the energy sector in order to create national competitive markets and integrate them into a unified European energy system. Legal bases for the development of foreign trade relations in the energy sector of the Russian Federation and the EU studied. It is established that the rules of the EU law, first of all the European Energy Charter and the Energy Charter Treaty, of Regulation I have a significant impact on the civil

A. O. Inshakova (✉)
Department of Civil and International Private Law, Institute of Law,
Volgograd State University, Volgograd, Russia
e-mail: gimchp@volsu.ru; ainshakova@list.ru

E. E. Frolova
Institute of State and Law, Russian Academy of Sciences, Moscow, Russia
e-mail: frolova_ee@rudn.university

I. P. Marchukov
Department of Civil and International Private Law, Volgograd State University,
Volgograd, Russia
e-mail: gimchp@volsu.ru

© Springer International Publishing AG, part of Springer Nature 2019
O. V. Inshakov et al. (eds.), *Energy Sector: A Systemic Analysis of Economy,
Foreign Trade and Legal Regulations*, Lecture Notes in Networks and Systems 44,
https://doi.org/10.1007/978-3-319-90966-0_13

regulation of foreign trade relations in the Russian Federation and provide the basis for its unification with EU legislation.

The unification of civil and legal regulation of the foreign trade turnover of the energy resources of the Russian Federation and, above all, of conflict regulation, continues to be a promising trend in its development. This evidenced by the changes analyzed in the work, introduced in 2013 in Sect. VI of Part Three of the Civil Code of the Russian Federation during the reform of the Civil Code of the Russian Federation.

The preservation of the trend toward unification with the EU right, at least for the medium term, also indicated by the results of the analysis of official documents that determine the priority directions for the development of civil and legal regulation in the energy sector in the Russian Federation.

The conclusion drawn, that the lag in the processes of unification of the national civil-law regulation of the foreign trade turnover of energy resources in the sphere of international cooperation with the EU jeopardizes the system of long-term contracts, the author believes, in the sphere of energy resources supply to the EU countries. Argued, that the extent of the application of the long-term contract system is also affected by excessive liberalization of the EU internal energy market. In particular, with regard to the preservation in such contracts of reservations about final destination points that limit the rights of buyers to resell Russian energy products within the EU. Noted, that the compulsory exclusion from reservations of long-term contracts of final destination points without unrestricted third party access to such a system will give gas importers advantages over gas exporters and entails unprofitable conditions for the fulfillment of obligations of a foreign trade contract on the part of the seller.

2 Materials

The reference in the work to the study of the general legal regulation of the foreign trade turnover of the EU energy resources is justified, first of all, by close foreign economic relations, in which the Russian Federation acts as a supplier. In addition, one should take into account the fact that the EU has a developed institutional and organizational structure and a high degree of economic integration of the regional association, which confirmed by the progressive creation within the EU of a single social, economic and legal space, including in the energy sector. In this regard, the role, institutional structure, process and results of the formation of a single energy economic and legal space in the EU, as well as the harmonization of Russian law with the European law were studied on the basis of the works of Abashidze and Inshakova (2016), Valde and Konoplyanik (2009), Voloshin (2004). This was facilitated by the EU regulatory framework, which includes such international acts as: the Treaty on the European Union (Maastricht, February 7, 1992) (as amended by the Lisbon Treaty in 2007), the European energy Charter, the Energy Charter

Treaty, which entered into force on 16.04.1998, the Final document of the Hague conference on the European energy Charter.

The pan-European conflict regulation and the issues applicable in the field of obligations arising from the foreign trade turnover of energy resources were also studied on the basis of secondary EU acts. Such as: Regulation No. 593/2008 of the European Parliament and the Council of the European Union "On the law subject to contractual obligations ("Rome I"), which replaced the Rome Convention on the law applicable to treaty obligations of 1980". Examination of the provisions of Regulation Rome I also made it possible to draw conclusions about the ongoing harmonization of the relevant rules in EU law with the provisions of the Civil Code of the Russian Federation. And to analyze the changes in Russian norms related to the solution of the conflict problem in the sphere of civil obligations complicated by a foreign element and harmonized with norms of Regulation Rome I.

For the study of the EU's overall energy policy was promoted by such strategic documents as: The Green Book of 2006, entitled "European Strategy for Sustainable, Competitive and Secure Energy"; Report of the Commission "Energy Policy for Europe"; The energy road map is 2050. The secondary documents studied in the course of the study and determining the EU's overall energy policy are the documents included in the Third Energy Package: Regulation No. 713/2009 on the establishment of the Agency for Cooperation of Regulatory Energy Bodies; Regulation No. 714/2009 on the conditions for access to networks for cross-border electricity exchanges; Regulation No. 715/2009 on the conditions for access to gas transmission networks. And—Directive 2009/72/EC, establishing general rules for the domestic electricity market; Directive 2009/73/EC, establishing general rules for the domestic natural gas market. A scientific vision of the problems of the EU's common energy policy was examined using the example of Dedenkulov (2015), Kaveshnikov (2013), Marchukov (2015a, b), Redkin (2008).

In order to study the prospects for the creation of a reliable integrated trans-European energy infrastructure of the EU, the following legal documents were studied: Regulation No. 347/2013 on guidance for the trans-European energy infrastructure of the EU, providing assistance in the implementation of certain infrastructural "projects of common interest"; Regulation No. 1391/2013 with an updated list of "projects of common interest"; Commission report "Long-term vision of infrastructure for Europe and beyond".

Research into the issues of the energy dialogue and harmonization of the external economic regulation of international relations between the Russian Federation and the EU in the energy sector was facilitated by the work of Bulaev (2013), Petrov (2013), Utkin (2014), Shumilo (2012), covering the economic and legal problems of implementing joint global energy projects, such as Nord Stream, Druzhba, and Blue Stream.

As legal documents that form the regulatory basis for foreign economic cooperation in the energy sector between the Russian Federation and the EU, the Partnership and Cooperation Agreement, the Memorandum on Industrial Cooperation in the Energy Sector between the Ministry of Fuel and Energy of the Russian Federation and the European Commission, The Roadmap for cooperation between Russia and the EU in the energy sector until 2050, the Memorandum on the mechanism of early warning in the energy sector within the framework of the Russia-EU Energy Dialogue.

Legal problems and priorities for the development of the legal regulation of the supply of energy resources of the Russian Federation to the EU have been highlighted taking into account the opinions expressed in the works of Ispolninov and Dvenadtsatova (2015), Konoplyanik (2005).

3 Methods

The author's study based on the universal scientific method of historical materialism. The general scientific methodological basis of the chapter was the methods of analysis, synthesis, generalization, as well as formal-logical, dialectical, system-structural methods.

The main scientific method, which made up the methodological basis of this part of the research, is the method of comparative legal analysis that made it possible to implement the scientific development of EU regulatory and legal acts that are part of the energy sector regulation system, including the turnover of energy resources. In addition, the applied method made it possible to compare the provisions of the EU legal acts not only with each other, but also compare them with the provisions of the current Russian legislation in the field of mandatory energy relations, in particular, the provisions of the Civil Code of the Russian Federation. Other methods of a private scientific nature also used: formal legal, the principle of evaluating legal processes, etc.

4 Introduction

The European Union (hereinafter referred to as the EU) has a significant influence on the external economic regulation of the world fuel and energy complex. The Russian Federation is not an exception. Moreover, as it was said in the 1st paragraph of Chap. 2, the majority of the counterparts of the largest Russian corporations engaged in cross-border management in the energy sphere identified by the study have the nationality of the member state of the EU. In addition, taking into account the conclusion made in the previous paragraph about the active participation of international integration associations in the process of forming the common energy policy of many countries, international lawmaking and unification of energy regulation through their bodies and institutions, referring to the study of the general legal regulation of the foreign trade turnover of EU energy resources logical and conditional. First of all, it is necessary to take into account the fact that the EU has passed through all stages of economic integration (Abashidze and Inshakova 2016) and has a developed institutional and organizational structure.

The target of creating the largest international association of European countries is regional integration through the formation of a single social, economic and legal space (Normative-legal acts 2017). The EU countries are the largest subjects of international trade relations, actively speaking on both sides of foreign trade contracts.

Energy issues have always been and given considerable attention as one of the most important factors in the development of European integration. Adequate legal regulation of relations in the energy sector is a necessary condition for the prosperity of the EU countries and the welfare of citizens.

5 The EU Legal System, Reflecting the Policy and Framework for Regulating the Energy Sector and the Turnover of Energy Resources

As already mentioned in Chap. 1, Paragraph 1, the European Energy Charter (Kolosov and Krivchikova 2007) is one of the main documents in the field of foreign policy not only on the scale of the EU. According to V. I. Voloshin, it was the European Energy Charter, that defined the main ways and principles of modern international energy cooperation (Voloshin 2004). In order to give legal effect to the main ideas set forth in the Charter, the Treaty on the Energy Charter was adopted, which came into force on 16.04.1998 (hereinafter—ECT).

Under Article 45, Russia applies the Treaty on an interim basis: "Each signatory agrees to temporarily apply this Treaty pending its entry into force for such a signatory in accordance with Article 44, to the extent that such provisional application does not contravene its constitution, laws or normative acts" (Valde and Konoplyanik 2009).

The ECT has a mixed subject composition. The ECT includes both states and international organizations (for example, the European Union). In addition, the ECT includes both importing countries and energy exporting countries. An analysis of the norms of the ECT on the functioning of the EU allows us to conclude, that there are provisions in the ECT, that directly regulate the energy sector (for example, Sect. XXI "Energy"). As well as norms that indirectly affect this area (for example, articles on consumer protection, the environment, fixing rules for the functioning of the external and internal market, special provisions for the functioning of the energy sector are contained in Sect. XVI "Trans-European Networks" of the ECT).

The ECT can't be called a static contract in connection with the continuous evolution of both the subject composition and content.

The content of the ECT covers the wide range of relations in the energy sector (Gudkov and Lakhno 2011) and its norms divided into three main components: trade, investment and transit. In the EU, conflict rules, that contain rules for determining the applicable law contained in Regulation No. 593/2008 of the European Parliament and the Council of the European Union, adopted in 2008, "On the law subject to treaty obligations ('Rome I')" (hereinafter referred to as the Regulation) (Higher School of Economics 2017), which intended to replace the Rome Convention on the Law Applicable to Contractual Obligations of 1980 (Convention on the Law Applicable to Contractual Obligations 1997). The Regulation has direct effect in the Member States of the EU and is legally binding. It covers the definition of the applicable law in the absence of agreement of the parties, the limits of autonomy of will and the definition of a treaty statute in the EU countries.

Answering the key questions, that are of particular importance for the normal functioning of the single EU market, and as successful example of unification of conflict regulation, the regulation provokes interest and the opportunity provided by it for the application of the law of a state that is not a member of the EU (potentially the rights of the Russian Federation).

According to the provisions of Presidential Decree No. 1108 of 18.07.2008 "On the Improvement of the Civil Code of the Russian Federation", on the approximation of "the provisions of the Civil Code of the Russian Federation with the rules for the regulation of the relevant relations in the law of the European Union" in the event of a change in Russian standards related to the resolution of the conflict problem in the sphere civil obligations, complicated by a foreign element, the norms of Regulation Rome I. The Regulation reflects modern trends in the development of conflict-of-law regulation of contractual obligations.

The preamble to this document recorded: "The normal functioning of the internal market requires that in order to increase the predictability of the outcome of litigation, to increase certainty with respect to the law to be enforced and to facilitate the free movement of judgments, the conflict of laws rules in the member-states indicate one and the same national law regardless of the country where the claim filed in court."

According to Article 3 of the Regulation, the contract is governed by the law chosen by the parties. The choice must be expressed or definitely follows from the provisions of the treaty or from the circumstances of the case. Consequently, the basis of the Regulation, as well as international private law in general, is the principle of autonomy of the will of the parties—the parties are free to choose the legal regulation applied to the concluded contract.

At the same time, the Rome Convention of 1980 placed first on the principle of the closest relationship, the Regulation establishes provisions for determining the law to be applied to specific treaties in the absence of agreement between the parties (art. 4): the contract of sale of goods is governed by the law of the country where the seller has his usual place of residence; the contract for the provision of services is governed by the law of the country where the service provider has its usual place of residence; the contract6 that has a real right to immovable property or a lease of immovable property is governed by the law of the country, where the immovable property is located; the contract on the sale of products is governed by the law of the country where the party selling the goods has its habitual residence, etc.

If the law, applicable to the contract is not possible to determine, taking into account the proposed in art. 4 of the Binding Regulations, or if the elements of the contract are subject to several bindings, the contract is governed by the law of the country where the usual place of residence of the party is located, which must perform the performance that is crucial for the content of the contract. If, from the circumstances of the case, it follows that the contract has clearly more close ties with another country than that indicated in paragraph 1 or 2, then the law of that other country applies (art. 4, paragraf 3). For cases in which the applicable law can't determined under paragraphs 1 and 2, the contract is governed by the law of the country with which it has the most close ties (art. 4, paragraf 4).

Article 20 of the Regulation excludes a return: the application of the law of a country means the application of the rules of law in force in that country, with the exception of the norms of its international private law.

Thus, the Regulation establishes a consistent application of law in the absence of choice: first defined by the principle of characteristic execution, and, if not possible, the right of closest connection. The new rules for determining the applicable law in the absence of a choice of parties established by the Rules differ from the Rome Convention of 1980, but are similar to the provisions of paragraph 1 of Article 1211 of the Civil Code of the Russian Federation (as amended by Federal Law No. 260-FZ of September 30, 2013).

However, the EU's generally consistent overall energy policy is reflected, as a rule, in the sources of "soft law", which serve as a basis for further adoption of legally binding acts.

The main documents reflecting the EU's modern energy policy include: The Green Paper of 2006 (The World of Energy Supply 2017) (Green paper)[1] under the name "European Strategy for Sustainable, Competitive and Reliable Energy" («A European Strategy for Sustainable, Competitive and Secure Energy») (Commission of the European Communities 2017) and a report from the Commission entitled "Energy policy for Europe" (Electronic Fund for Legal and Regulatory Documentation 2017).

The green book contains a description of the situation and problems in the energy sector. The document has three key objectives of energy development (ensuring sustainable development, competitiveness and reliability of supply). In addition, it defines the priority areas for the energy policy (internal energy market ensuring reliability of energy supplies, a unified approach to combating climate change; completion of construction of a single gas and electricity market; development of innovative activities; maintenance of reliability and competition in the sphere of energy supply; way to an agreed foreign policy). Each part of the Green Book is devoted to the corresponding priority area and contains a list of measures necessary to implement the measures (Abashidze and Inshakova 2016; Gudkov 2006).

---

[1]Note: The Green Book is a document published by the Commission in the sphere of one of the policies implemented by the EU, which intended for discussion. First of all, the document is addressed to interested parties—organizations and individuals, who are involved in consultations and debates.

The Commission's message "Energy Policy for Europe" is a strategic review of the energy situation in Europe.

The first part of the document is devoted to the main problems in the energy sector, among them: competitiveness, sustainable development and security of supply.

The second part indicates the strategic objective of the EU energy policy: reducing dependence on imported carbons; combating climate change; development and employment growth, which should ensure the reliability and availability of energy for consumers (Marchukov 2015a, b).

The Commission's message "Energy Policy for Europe" (hereinafter referred to as the Communication) is included in the third energy package submitted by the Commission on September 19, 2007 and defining the postulates of the EU's common energy policy. The main objective of the third energy package is the creation of an open and honest energy market, the unification of the rather dissociated energy market of the EU. Innovations had to destroy national barriers to trade in electricity and natural gas by increasing security of supply and developing competition at the EU level, which would provide consumers with a wider choice and improve the quality of services provided (RIA Novosti 2017).

In addition to the Communication, the Third Energy Package includes three regulations: Regulation No. 713/2009 on the establishment of the Agency for Cooperation of Regulatory Energy Bodies, Regulation No. 714/2009 on the conditions for access to networks for cross-border electricity exchanges, Regulation No. 715/2009 on conditions for access to gas transmission networks, and two directives: Directive 2009/72/EC establishing general rules for the domestic electricity market and Directive 2009/73/EC establishing general rules for the domestic natural gas market.

These acts entered into force on March 1, 2011, and they include a number of innovations related to: (1) the separation of vertically integrated companies in the most brutal form, namely, property separation; (2) strengthening the independence of regulatory bodies through their isolation, both from industry and from state apparatus; (3) improve the scheme of interaction between operators of transportation systems and ensure the optimal management of these systems, including through the creation of European networks of operators of gas and electricity transmission systems; (4) improve of the system of consumer rights protection, development of the Energy Charter of Consumers, as well as the functioning of the Citizens' Energy Forum; (5) strengthening the unity of the member states with a view to bringing together the national markets, as well as providing mutual assistance in the event of difficulties with the supply of energy resources.

In general, the third energy package is aimed at eliminating the shortcomings of the current legislation by introducing new rules for regulating relations in the energy sector. All the proposed changes are an evolution of earlier ideas and indicate their continuity.

In December 2011, the Commission adopted the Energy Roadmap-2050, which contains a wide range of measures in various areas of the EU's energy policy (European Social and Economic Committee 2016).

6   The Harmonized Energy Policy of the EU as an Integral Element of a Single Economic Space

Recently, the issue of energy security in the energy policy of the EU has become important (Kaveshnikov 2013). This may be due to external (primarily Ukrainian gas crises), and internal factors (large-scale expansion of the EU in 2004–2013). In the EU, the priorities of the energy policy shift towards an external dimension and an infrastructure track (Guarnieri 2016).

Approved in April 2013, Regulation No. 347/2013 on the management of the trans-European energy infrastructure of the EU provides for assistance in the implementation of certain infrastructural "projects of common interest." On October 14, 2013, the European Commission announced plans for the development of the energy infrastructure: Regulation No. 1391/2013 with an updated list of "projects of common interest" and "Long-term vision of infrastructure for Europe and beyond" according to which "an adequate, integrated and reliable energy infrastructure is a critical prerequisite for achieving the goals of not only the energy policy but also the economic policy of the Union" (MGIMO University 2016).

The Commission plays the main role in the implementation of the EU energy policy.

The functions of the Commission in the field of energy mainly concentrated on coordination, supervision and control of the activities of participants in the energy market of the Union. For example, Regulation No. 736/96 of 22 April 1996 on notification to the Commission of investment projects affecting the interests of the EU in the oil, gas and electricity sectors, obliges to disclose information on investors on major investment projects in the energy sector to the Commission (Dolega 2017).

The Commission and EU member states actively use such form of acts as strategies: the Strategy for Biological Fuels "An EU Strategy for Biofuels", 2006, "Energy 2020 Strategy for Competitive, Sustainable and Reliable Energy".

In January 2014, the European Commission published a program of the EU strategy on energy and combating climate change until 2030, within which three main objectives are planned: increasing the share of renewable energy in the structure of energy consumption; reduction of greenhouse gas emissions; the resumption of activities aimed at improving energy efficiency, the development and implementation of new indicators and the necessary changes in the management system to ensure the competitiveness and security of the EU energy system (Foreign Policy 2017).

The EU's energy security strategy, published by the Commission in May 2014, is aimed at maintaining price stability and ensuring the continuity of energy supplies. The Strategy specifies short-term, medium-term and long-term measures from the development of energy infrastructure and renewable energy sources, the increase of gas reserves to energy efficiency, the diversification of routes and sources of supply, building up domestic energy production, completing the formation of the domestic energy market and implementing the concept of a unified voice in the foreign energy policy (Dedenkulov 2015).

Thus, the EU is actively pursuing a policy of unifying legal regulation in the energy sector in order to create national competitive markets and integrate them into a unified European energy system (Redkin 2008).

The main role in the formation of integrated legal regulation of the energy sector, including in the area of trade in energy resources, belongs to the soft law of the EU. The last 10 years are a confirmation and clearly show that the development of energy in the EU determined by various kinds of programs. Only in the field of renewable energy in the EU's total energy consumption they are accepted several: «FP IV», «FP V» (Marchukov 2015a, b).

Among the acts of this kind of soft law are: strategies, programs, plans, green and white books, recommendations, memoranda, framework decisions of the Council, etc. Acts of soft law of the EU are an important element of the system of program guidelines that regulate the energy sector, which, despite the non-binding nature, predetermine the formation of the EU's future right and strategically outline the framework for rulemaking (Marchukov 2015a, b). With the help of soft law acts, the priority areas of energy activities are substantiated, as well as the main measures for their implementation, as a result of which the Commission receives a mandate to develop uniform norms legally binding for all actors and throughout the EU. The legal norms elaborated in this way replace the acts of soft law after the necessary degree of readiness of the Member States for their acceptance reached.

Thus, all of the above allows us to recognize the harmonized energy policy of the European Union as an integral element of a single economic space. In this context, EU energy law is an example of a liberal approach to regulating regional energy markets. Thanks to the system approach, the legal rules for the functioning of the internal energy market of the EU can taken, as a model for industry associations within the CIS and the EEU. However, the supranational way of its formation and the multiplicity of program and strategic framework acts, which serve as a basis for further development of unified rules through mandatory rules, do not fully correspond to the objectives of Eurasian integration with the participation of the Russian Federation, the energy aspects of which should be subject to the principle of mutual benefit and flexible regulation in accordance with the method of harmonization of law.

7 Harmonization of Foreign Economic Regulation of International Relations Between Russia and the EU in the Energy Sphere

As for the relationship between the EU and Russia, it should remembered that the European market has always been the main direction of export of national oil products. In the energy sector, the EU is the largest importer of oil, gas, uranium, coal and other fuel resources supplied from the Russian Federation. The Russian Federation exports energy products to virtually all European countries. According to official statistics in 2015, oil exports from Russia amounted to 244 million tons, worth \$89.6 billion. Russia supplies about 36% of the total volume of all oil that enters the EU (Technoblog 2017). Supplies of gas from Russia in 2016 were equal to 161.5 billion (On the line 2017).

In recent years, Russian and European companies have implemented several global energy projects. Among them are Nord Stream, Druzhba and Blue Stream, which are the largest gas mains between Russia and Germany (Bulaev 2013). A gas pipeline project, South Stream, is the key for the international energy markets.[2] However, due to the worsening of political relations between Russia and the EU in 2014–2015, President of the Russian Federation V. V. Putin announced the freeze of the project for an indefinite period.

The high degree of mutual economic interest, the strategic nature of international relations between Russia and the EU countries, and the exacerbation of foreign policy make it necessary to conduct research in the field of legal regulation of energy cooperation between the EU and Russia.

Currently, the main form of foreign economic cooperation between the EU and the Russian Federation is the energy dialogue, whose goal is to build harmonious and long-term relationships in the field of trade in energy resources. The subject of the dialogue is the oil, natural gas, coal and nuclear materials market.

The beginning of cooperation in the form of dialogue is associated with the international EU-Russia Summit, held in 2000 in Paris. The initiative of foreign economic cooperation with Russia came from the Chairman of the EU Commissions R. Prodi, who proposed that Russia strengthen and develop energy partnerships by increasing energy supplies to the countries of the Union in exchange for investment and new technologies.

The first international normative legal act, which consolidated the course of the EU and Russia for mutual integration and the formation of a single energy space, was the Partnership and Cooperation Agreement (hereinafter—the PCA) (EU portal 2015a). The PCA concluded for a period of 10 years with the subsequent annual automatic renewal of the Agreement, if neither side declares its denunciation. The Union of Right Forces laid the foundation for the development of foreign economic relations between Russia and the EU, including in the energy sector. So, the most important provisions of the Agreement are fixed in the section "Trade in Goods", which defines the basic principles of trade between Russia and the EU countries, among which one can single out the regime of the favored nation, the avoidance of direct or indirect discriminatory taxation, freedom of transit, the possibility of protecting the domestic market with the use of anti-dumping methods. Also, PCA regulates the issues of cash flow between market participants. In accordance with

---

[2]An unrealized international gas pipeline project that allows diversifying Russian natural gas supplies to Europe and reducing the dependence of suppliers and buyers on unreliable transit countries (*Note by the authors*).

Article 52 of the Agreement, the parties undertake to allow the conduct in a freely convertible currency of any current payments between residents (EU portal 2015a). Thus, the PCA establishes the principle of free flow of funds and capital liberalization, including those related to portfolio investments, commercial loans and loans.

A significant role in the process of harmonizing the external economic and legal regulation of Russia's relations with the EU in the energy sector is played by Article 55 of the Agreement, according to which the parties recognize that an important condition for strengthening economic ties between Russia and the Community is the approximation of legislations (EU portal 2015a). This process affects many areas, including energy. As P. S. Shumilo rightly believes, it was noisy that such rapprochement is necessary to remove the financial and legislative barriers that hinder the creation of a free economic zone for the energy market and reduce the costs of the EU countries for the transit of energy resources supplied from Russia (Shumilo 2012). Priority directions of development of foreign economic relations of Russia and the EU in the field of energy named in Article 65 of the PCA. Among them: improving the quality and safety of energy supply; development of a unified energy policy; improvement of regulation of the energy market that meets the conditions of a modern market economy; modernization of energy infrastructure; management and technical training in the energy sector, etc. (EU portal 2015a). Also, Article 65 of the PCA establishes the crucial provision that cooperation between Russia and the EU in the field of energy is carried out on the basis of the principles of a market economy and the European Energy Charter on the basis of the gradual integration of energy markets in Europe (EU portal 2015a).

Based on the foregoing, it can be concluded that at present the Partnership and Cooperation Agreement is the fundamental legal basis for the development of relations between the EU and Russia. In recent years, the leadership of the EU and Russia are actively discussing the need for a new Partnership and Cooperation Agreement in place to replace the current Agreement. However, while Russia and the EU are not ready to take the next step towards the rapprochement of political and economic interests, in spite of the fact that the need to adopt a new Agreement exists and caused primarily by the need to improve the legal regulation of the energy market. Energy cooperation, which due to its scale, according to experts, was to become an anchor and locomotive of bilateral cooperation, constantly generates mutual phobias of "dependence", "discrimination", turning into a zone of constant tension. For detente, it will be necessary to adopt a new Agreement with the unification of the standards of industrial energy cooperation. However, the current geopolitical situation, and above all the political crisis in Ukraine and Russia's participation in the fighting in Syria, led to an aggravation of economic relations between the EU countries and the Russian Federation, which resulted in the suspension of negotiations on the adoption of a new PCA (Utkin 2014).

In order to harmonize the external economic regulation of international relations between Russia and the EU in the field of energy, on February 11, 1999, a Memorandum on industrial cooperation in the energy sector was signed between the Ministry of Fuel and Energy of the Russian Federation and the European

Commission (EU portal 2015a). The significance of this document is to create and develop a leasing fund financed from various commercial and institutional sources, which should created to provide leasing financing to energy sector enterprises in Russia and the EU. As K. E. Petrov, an established system of leasing in the energy market between Russia and the EU, would expand the freedom of energy trade and weaken the existing contradictions in the energy sphere (Petrov 2013).

At present, the main document regulating foreign economic relations between Russia and the EU in the energy sector is the Roadmap for Cooperation, signed on March 22, 2013 (EU portal 2015b). This document is an agreement between Russia and the Union on the main directions, objectives and stages of cooperation until 2050 in the field of trade in gas, oil, electricity and renewable energy sources. V. P. Gerich and C. Clatings note that the Roadmap aimed at studying various scenarios and their possible impact on EU-Russia energy relations, including their implications for different energy sectors. Long-term opportunities and risks for the overall situation with demand and supply should studied. The potential for long-term cooperation in the energy sector should be explored, taking into account the creation of the necessary infrastructure, investment conditions, convergence of the regulatory framework and the development of an efficient and open to innovation energy sector (Ministry of Energy RF 2015). The "road map" is the official regulator of foreign economic relations between Russia and the EU in the energy sector. The final target of this agreement is the formation by 2050 of a single free economic space in the energy market on the scale of the Eurasian continent. The document contains a plan for developing foreign economic relations of the Parties separately for each energy sector: electricity, gas, oil, renewable resources.

The Road Map also contains an indication of a number of Russian energy companies that play a key role in the process of forming a free economic space. In the electric power industry this is: PJSC RusHydro, JSC Rosenergoatom, OAO RAO UES of Russia. In the gas and oil industry: PJSC Gazprom, PJSC Lukoil, PJSC NK Rosneft. These companies are the main energy exporters to the EU countries, with the adoption of the Road Map, the nature of the supplies ordered by them in accordance with the objectives of this agreement.

One of the targets of the Roadmap for Co-operation is the harmonization of the legislation of Russia and the EU for the optimal development of market relations in the energy sector and the formation of favorable conditions for investment. The urgency of this target, that at present there are no international legal acts adopted and ratified by the parties that regulate the practical issues of energy trade in the EU-Russia system.

Thus, the Roadmap is intended to strengthen the systemic interaction between Russia and the EU in the field of energy, through the improvement of the regulatory and legal framework governing energy relations, laying a solid foundation for the gradual convergence of rules, standards and markets in the field of energy, which in turn will ensure an increase in the volume of mutual investment and the exchange of technology (EU portal 2015b).

When carrying out energy exports to the EU countries, Russian organizations adhere to the WTO Trade Rules, as well as the international Incoterms-2010 rules relating to the transport and supply of goods (Incoterms 2010). However, to settle trade issues at the current stage of the energy market development, this regulatory framework is not enough (Austvik and Lembo 2017). The adoption by the EU countries of the "Third Energy Package", which is a set of directives in the sphere of energy supplies, has caused the growing contradictions between Russia and the Union because of the infringement of Russia's economic interests in the energy sphere.

Another important document adopted to harmonize the external economic regulation of international relations between the Russian Federation and the EU in the energy sector is the Memorandum on the Mechanism of Early Warning in the Sphere of Energy within the framework of the Russia-EU Energy Dialogue (EU portal 2015c). This document was signed on November 16, 2009 in Moscow, its purpose is to determine the procedure for implementing joint measures for the prevention and prompt response in case of occurrence or threat of an emergency. The need to create such an agreement arose as a result of the gas conflict between Russia and Ukraine in 2009, which was a dispute over the debts, prices, supplies and transit of natural gas.

The objectives of signing the Memorandum are: to ensure a stable and unhindered energy supply, to prevent and overcome with minimal negative consequences of emergency situations in the energy sphere. The scope of the Memorandum includes situations associated with an exceptionally high growth in demand for energy products, when Russian companies are forced to make urgent and large deliveries without any additional monetary compensation from importers; interruption of energy supplies due to the aggravation of the political situation in the region and in other cases. It should be noted, that this document does not contain any sanctions against third parties, but involves concerted joint actions to prevent the cessation of supplies, by including a certain procedure, initiated by either party.

The formation of a single European energy market continues, therefore, the legal regulation of energy in the EU includes unified norms, binding for the member countries, and in some cases—national legal regimes.

The cooperation of the Russian Federation with the united Europe is, in essence, complex and includes cooperation both with the European Union as a whole and with its individual member states.

The specifics of the legal regulation of energy relations between Russia and the EU, determined by the totality of the norms of the domestic legislation of Russia and the EU, as well as the domestic legislation of the countries participating in the integration regional association, allows us to speak about the presence of a pan-European and national levels of legal regulation in the field under study. Such a conglomerate of normative interaction of supranational and national binding regulation in the field of energy in practice allows for more flexible planning of contractual relations with the EU member states, joint investment projects and projects in the field of services.

An analysis of the energy regulation of the EU and the Russian Federation made it possible to talk about the coincidence of the parties' priorities for the medium term. First of all, they include issues of ensuring energy efficiency, energy security, stability of energy supplies. In the context of a long-term perspective, the issue of diversification of energy sources becomes particularly important.

8 The Unified Energy Market of Europe and the Priorities for the Development of the Legal Regulation of the Supply of Energy Resources of the Russian Federation

Regarding energy regulation in the sphere of foreign trade of energy resources, the main task of the EU within the framework of the concept of creating a single energy market in Europe while reducing the politicization of the "energy issue" is to reduce the region's dependence on direct supplies from Russia (European External Action Service 2017).

In this regard, the issues of the unification of the legal regulation of the foreign trade turnover of energy resources with integrated Europe are quite acute for Russia. Legal regulation of the foreign trade turnover of Russia's energy resources as an exporting country should take into account the main priorities for the medium and long term—technological modernization of the country's fuel and energy complex, as well as the development of international cooperation within organizations and integration associations. Lagging the development of legal regulation of foreign trade turnover in the energy sector under these two conceptual directions jeopardizes the system of long-term contracts that are predominant in the sphere of supplying Russia's energy resources to the EU countries.

Currently, the process of liberalization of the internal energy market within the EU raises questions about the extent of the application of the long-term contract system, in particular, the retention of such clauses on final destination points restricting the rights of buyers to resell Russian energy products within the EU.

Disclaimers on final destination points (or territorial restrictions on sales) allow gas suppliers to sell it to various buyers (whose end-users are located at various distances from the delivery points where buyers purchase this gas) at different prices and on different terms on the same point of acceptance. Such disclaimers are forbidden to the buyer acquiring Russian gas for its delivery, for example, to Austria, to resell it in France (to France) or to the Czech Republic (in the Czech Republic). And, thus, to receive unjust enrichment, due to the availability of different prices, set by one supplier in the same geographical location for deliveries to various points of final destination (Konoplyanik 2005).

Thus, the problem of the existence of long-term contracts for the supply of Gazprom gas with final destination clauses is understood by the European Commission as Gazprom's desire to set obstacles to free trade in gas between the EU member states, which makes it impossible for the free and unified gas market to function effectively in the Europe.

The Commission, through the initiation of antitrust investigations, decided to use very strict methods to stop the practice of including in such contracts restrictive competition the disclaimers (Ispolninov and Dvenadtsatova 2015). So, in 2003–2004, under strong pressure from the Commission, significant changes were made to Gazprom's two large Russian contracts for the supply of gas to the EU countries related to reservations about final destination points: agreements with the Italian company ENI and the Austrian company OMV.

On March 13, 2017, it became known from the European Commission (Press Release 2017) that Gazprom agreed to exclude from the contracts with European buyers of Russian gas conditions that forbid European buyers of gas supplied by Gazprom to resell it to other EU countries. This decision is conditioned by two circumstances: (1) to prove readiness to follow the rules of the antimonopoly law and the Third Energy Package of the European gas market; (2) to settle the anti-monopoly investigation initiated against the corporation in 2012. Gazprom was accused of trying to divide the European gas market. In particular, the represen-tatives of the European Commission stated that "Gazprom" prevents diversification of gas supplies to Central and Eastern Europe, and establishes unfair prices (RBC-news, shares, exchange rates 2017).

On the one hand, the presence of reservations about final destinations reduces the possibility of cross-border trade, initially reducing the number of possible gas suppliers to end users, hampers the creation of a free and unified gas market in the EU. On the other hand, excluding such clauses from foreign trade long-term con-tracts without unrestricted third-party access to such a system will give gas importers advantages over gas exporters. In this connection, the search for a compromise and effective decision on this issue in the current international treaty practice is an actual task.

9 Conclusion

In conclusion, it should noted that until very recently the European Union was regarded in the Russian doctrine as the most successful model of economic inte-gration of states, in turn, the EU right as an exemplary regulatory system for the unification of national legislation in various areas of legal regulation. This view of domestic researchers was largely due to the large volume of trade between the EU and Russia and the expansion of cooperation in various areas, including in the energy sphere. However, the foreign policy situation in the EU in recent years has had a significant impact on its former assessment, overly idealizing cohesion, federation of unity and economic unity (Abashidze and Inshakova 2016).

Given the above circumstances, the processes of economic integration of Russia with the EU in the development, development and supply of fuel and energy resources, as well as the unification of energy regulation can drastically change the vector of its development (Marchukov 2015a, b).

Today we can't longer confidently state the need for a strategic partnership agreement in the near future, which will allow us to formalize the solution of many politicized issues in order to achieve the stability of Russian-European energy relations.

The necessity for the signing of the Treaty about Energy Charter by Russia is also doubtful. Despite the fact that this document provides favorable conditions for long-term contracts preferential for the Russian side and gives them the status of international treaties for the states that signed the Treaty about Energy Charter. This need was more appropriate in the context of building a single energy market and a pan-European economic space as a whole.

However, we believe that it is necessary to continue work on the approximation of legal regulation in the energy sphere of the EU and Russia, the creation of an attractive investment climate that allows foreign investors to be active, in the projects of the transport infrastructure of the oil and gas industry, and the integration of electricity markets. Unification of legal regulation of foreign trade turnover of energy resources in Russia and the EU, first of all, technical standards, legislation in the field of energy efficiency will allow laying a solid foundation for the gradual approximation of rules, standards and markets in the energy sector, which will increase investment and technology exchange.

The development of a unified legal basis for intercountry cooperation between the Russian Federation and the EU can be the development, creation and adoption of bilateral international agreements at the intergovernmental level.

As a final conclusion, it can be stated that the analysis of civil-law regulation of the energy sector within the EU and its priority areas of development has shown, that the international regional association does not depart from the course taken a few decades ago to gradually unify legal regulation in the energy sphere with the aim of creating national competitive markets and their integration into a unified European energy system.

Norms of EU law, primarily the European Energy Charter and the Treaty about Energy Charter, Regulation I Rome, have a significant impact on the civil regulation of foreign trade energy relations in the Russian Federation and provide the basis for the unification of the legal regulation of Russian civil legislation with EU legislation. In this regard, the unification of the legal regulation of the foreign trade turnover of energy resources of the Russian Federation with integrated Europe continues to be a promising trend in the development of legal regulation of the foreign trade turnover of energy resources and, above all, the conflict. This is evidenced by the changes introduced in 2013 during the reform of the Civil Code of the Russian Federation in Sect. VI of Part Three of the Civil Code of the Russian Federation and analyzed in the monograph. Among them are the norms of the Civil Code of the Russian Federation unified with Regulation I: item 1 of article 1211 of the Civil Code of the Russian Federation with regard to limiting the use of the law of close communication in the resolution of the conflict issue and giving priority to the definition of a treaty statute in the absence of an agreement between the parties on the applicable law of the theory of characteristic execution (Article 4 of Regulation Rome I). And also regarding the solution of the problem of mobile conflict (Article 19 (3) of the Rome I Regulation); disclaimer 3 of Article 1210 of the Civil Code of the Russian Federation, fixing the reservation to limit the retroactive force of the subsequent choice of the applicable law similar to that stipulated in Article 3 (2) of the Rome I Regulation; a new version of clause 9 of Article 1211 of the Civil Code of the Russian Federation, introducing a corrective clause (Article 4, paragraph 3 of Regulation I Rome); introduction to article 1222.1 of the Civil Code of the Russian Federation of the institution of pre-contractual liability (Article 12 of Regulation I Rome).

The preservation of the trend toward unification with the EU right, at least for the medium term, is also spoken by the results of the analysis of official documents that determine the priority directions for the development of Russian energy regulation. As an example, we can cite the document "Conceptual Approach to the New Legal Framework for International Cooperation in the Sphere of Energy (Targets and Principles)" published on the official website of the President of Russia, based on the principles of the Energy Charter (Kolosov and Krivchikova 2007).

The recent decision of the largest gas exporter to the European market of Gazprom to exclude from the contracts with European buyers of Russian gas reservations about final destination points that limit the rights of buyers to resell Russian energy products within the EU is a confirmation of the unification of law in the EU and the Russian Federation. Such a decision demonstrates readiness to follow the rules of EU antitrust law and the EU Third Energy Package.

Lagging the process of unification of the national civil-law regulation of the foreign trade turnover of energy resources in the sphere of international cooperation with the EU jeopardizes the system of long-term contracts that are predominant in the sphere of supply of Russia's energy resources to the EU countries.

It affects the scale of the application of the long-term contract system and the excessive liberalization of the internal energy market of the EU, in particular with regard to retaining in such contracts clauses on final destination points that limit the rights of buyers to resell Russian energy products within the EU. The compulsory exclusion from reservations of long-term contracts of final destination points without unrestricted third-party access to such a system will provide gas importers with advantages over gas exporters and entails unprofitable conditions for the fulfillment of obligations of the seller's foreign trade contract.

References

Abashidze, A. H., & Inshakova, A. O. (2016). *European Union law: A textbook and a workshop for undergraduate and graduate studies*. Moscow: Yurayt.

Austvik, O.G., & Lembo, C. (2017). EU-Russian gas trade and the shortcomings of international law. *Journal of World Trade, 51*(4), 645–674.

Bulaev, S. A. (2013). South stream gas pipeline: Challenges, innovations, prospects. *Bulletin of Kazan State University, 5,* 34–36.

Commission of the European Communities. (2017). *European Strategy for Sustainable, Competitive and Reliable Energy*. Retrieved January 22, 2017, from http://journal.esco.co.ua/2011_4/art148.pdf.

Convention on the Law Applicable to Contractual Obligations (concluded in Rome on June 19, 1980). (1997). United Nations, New York.

Dedenkulov, A.V. (2015). European Union: Evolution of energy policy priorities. *Modern Europe, 1*(61), 116–125.

Dolega, W. (2017). Selected aspects of realization of strategic energy infrastructural investments. *Rynek Energy, 130*(3), 3–8.

Electronic Fund for Legal and Regulatory Documentation. (2017). *Commission Communication «Energy Policy for Europe»*. Retrieved January 22, 2017, from http://docs.cntd.ru/document/902168991.

EU portal. (2015a). *Memorandum on industrial cooperation in the energy sector.* Retrieved November 22, 2015, from http://www.russianmission.eu/userfiles/file/partnership_and_cooperation_agreement_1997_eng.pdf.

EU portal. (2015b). *Roadmap for cooperation between Russia and the EU in the energy sector until 2050.* Retrieved November 20, 2015, from http://www.russianmission.eu/sites/default/files/user/Roadmap%20RussiaEU%20Energy%20Cooperation%20until%202050_eng.pdf.

EU portal (2015c). *Memorandum on the Early Warning Mechanism.* Retrieved November 20, 2015, from http://www.russianmission.eu/userfiles/file/memorandum_on_early_warning_mechanism_2009_eng.pdf.

European Commission-Press Release. (2017). *Antimonopoly: The Commission welcomes comments on Gazprom's commitments regarding the Central and Eastern European gas markets.* Retrieved March 23, 2017, from http://europa.eu/rapid/press-release_IP-17-555_en.htm.

European External Action Service. (2017). *Pierr N how to deal with Russian gas. ECFR. EU.* Retrieved January 12, 2017, from http://eeas.europa.eu/russia/docs/eu_russia_progress_report_2010_en.pdf.

European Social and Economic Committee. (2016). *Energy road map until 2050.* Retrieved December 17, 2016, from http://www.eesc.europa.eu/resources/docs/energy-roadmap-2050.pdf.

Foreign Policy. (2017). *Foreign policy-research moscow organization for foreign policy.* Retrieved January 22, 2017, from http://www.foreignpolicy.ru/analyses/energeticheskaya-strategiya-es-do-2030-goda/.

Guarnieri, M.A. (2016). Lesson from past energy crises. *IEEE Industrial Electronics Magazine, 10*(2), 59–63.

Gudkov, I. V., & Lakhno, P. G. (2011). Legal regulation of international energy relations: State and development prospects. *Business Law, 2,* 29–42.

Gudkov, I. V. (2006). Building an energy security system: The EU green paper on energy policy. *Oil, Gas and Law, 2,* 47–54.

Higher School of Economics. (2017). *Regulation no. 593/2008 of the European parliament and the council of the European Union «On the law subject to application to contractual obligations (Rome I»).* Retrieved March 24, 2017, from https://pravo.hse.ru/intrilaw/doc/040001.

Incoterms. (2010). Publication of ICC, 715 2011, Infotropic, Moscow.

Ispolninov, A. S., & Dvenadtsatova, T. I. (2015). Long-term gas contracts in EU practice: Prospects for the future or future without prospects? *Law, 1,* 96–106.

Kaveshnikov, N. U. (2013). Development of the external energy policy of the European Union. *Bulletin of MGIMO University, 4*(31), 82–91.

Kolosov, Y. M., & Krivchikova, E. S. (2007). *Current international law.* Moscow: Yurayt, International Relations.

Konoplyanik, A. A. (2005). Russian gas for Europe: On the evolution of contractual structures (from long-term contracts, sales at the border and reservations about final destination to other forms of contractual relations? *Oil, Gas and Law, 3,* 33–44.

Marchukov, I.P. (2015a). Advantages of legal forms of fixing the basic foundations of the EU energy policy. *Bulletin of the VolSU. Series 5. Jurisprudence, 3*(28), 115–119.

Marchukov, I.P. (2015b). Common Interests of the BRICS countries in the energy sector: Problems of the common energy policy and energy law. *Bulletin of the VolSU. Series 5. Jurisprudence, 4*(29), 167–174.

MGIMO University. (2016). *Kaveshnikov NU «I love our plans gromadyo»: The European Union plans to develop energy infrastructure.* Retrieved December 22, 2016, from http://old.mgimo.ru/news/experts/document247540.phtml.

Ministry of Energy RF. (2015). *«Roadmap» for EU-Russia energy cooperation: Interim report 2011.* Retrieved November 22, 2015, from http://minenergo.gov.ru/activity/co-operation/russia_eu/road_map/.

Normative-legal acts. (2017). *The treaty on the European Union (Maastricht, February 7, 1992) (in the edition of the lisbon treaty of 2007).* Retrieved January 22, 2017, from http://ppt.ru/newstext.phtml?id=25293.

On the line. (2017). *How much gas does Gazprom sell to Europe?.* Retrieved January 12, 2017, from https://www.nalin.ru/skolko-gaza-prodaet-gazprom-v-evropu-1480.

Petrov, K.E. (2013). The present and future of the Russia-EU energy dialogue (development of the EU-Russia energy dialogue in 2012). *Bulletin of MGIMO University, 3*(30), 7–10.

RBC-news, shares, exchange rates. (2017). *Gazprom agreed to lift restrictions on the re-export of gas to Europe.* Retrieved March 23, 2017, from http://www.rbc.ru/business/13/03/2017/58c685899a7947c9e0693c75.

Redkin, I. (2008). Legal regulation of energy in Russia in conditions of globalization. *Proceedings of the Institute of State and Law of the Russian Academy of Sciences, 4,* 64–90.

RIA Novosti. (2017). *The third energy package of the EU and Russia's attitude towards it.* Retrieved January 27, 2017, from https://ria.ru/spravka/20130530/933499962.html.

Shumilo, P. S. (2012). Energy dialogue between Russia and the European Union-the basis of European energy security. *Russian Foreign Economic Bulletin, 4,* 8–11.

Technoblog. (2017). *Media: Russia topped the Top 10 oil suppliers in the EU.* Retrieved January 12, 2017, from http://teknoblog.ru/2016/08/13/67062.

The World of Energy Supply. (2017). *Green Book. European energy security strategy.* Retrieved January 22, 2017, from http://esco.co.ua/journal/2006_7/art063.pdf.

Utkin, S. A. (2014). Russia-EU energy dialogue: Development prospects. *Bulletin of MGIMO, 1* (57), 8–12.

Valde, T., & Konoplyanik, A. (2009). The energy charter treaty and its role in the world energy industry. *Oil, Gas and Law, 1,* 46–50.

Voloshin, V.I. (2004). *EU-Russia energy dialogue,* Russian-European Center for Economic Policy (RECEP). Moscow.

Common Energy Policy and Development Paths Legal Regulation of Foreign Trade Turnover of Energy Resources in the CIS and in the EEU

Agnessa O. Inshakova, Alexander I. Goncharov
and Igor P. Marchukov

1 Annotation

The authors studied the documents, within the framework of which the interaction of the CIS member countries in the energy sphere is currently being implemented. Due to the prevailing economic ties in the energy sector, the legal harmonization, unification and coordination of the legislations of the CIS member states are necessary measures not only in the sense of building a common legal space for the Commonwealth countries, but also for the purpose of convergence with the norms of international law. The problem of choosing the law applicable to the contract between subjects of trade turnover of different CIS countries is considered.

The relevance of the study of sources of legal regulation of foreign trade turnover in the EEU member states in this monograph is due to the fact, that in the next decade the EEU member countries will become key partners of the Russian Federation in the field of trade in oil and gas resources. Since by 2025 the countries of the Eurasian Economic Union intend to move to a common oil and gas market and a common economic space that allows increasing the volume of exports and imports of energy resources by several times.

A. O. Inshakova (✉) · A. I. Goncharov
Department of Civil and International Private Law, Institute of Law,
Volgograd State University, Volgograd, Russia
e-mail: gimchp@volsu.ru; ainshakova@list.ru

A. I. Goncharov
e-mail: gimchp@volsu.ru; goncharova.sofia@gmail.com

I. P. Marchukov
Department of Civil and International Private Law, Volgograd State University,
Volgograd, Russia
e-mail: gimchp@volsu.ru

© Springer International Publishing AG, part of Springer Nature 2019
O. V. Inshakov et al. (eds.), *Energy Sector: A Systemic Analysis of Economy,
Foreign Trade and Legal Regulations*, Lecture Notes in Networks and Systems 44,
https://doi.org/10.1007/978-3-319-90966-0_14

The CIS and the EEU are international organizations that have united the countries of the post-Soviet space that have similar legislation and a vision of international politic. The analysis of the legal regulation of the foreign trade turnover, of energy resources of the member countries of regional integration associations, based on the example of the Republic of Armenia and the Republic of Kazakhstan, confirmed the use of legal mechanisms similar to Russian law. Law, which including the structure of civil legislation, as well as, approaches to contractual regulation and issues of applicable law, which is due to the previously achieved uniformity in the post-Soviet legal space.

An analysis of the priority areas of the energy policy of the CIS and the EEU in the energy sector led to the conclusion that in the process of creating a common energy regulation of the CIS member countries, emphasis placed on harmonizing it, by developing program framework acts. Regarding the EEU, it is justified that a more effective method will be to unify norms by concluding bilateral and multilateral agreements and developing legally binding standards.

2 Materials

The authors study the energy policy and the main directions for the development of the legal regulation of the non-trade turnover of energy resources in the Commonwealth of Independent States and the Eurasian Economic Union on the basis of a study of the current regulatory framework. The analyzed array of regulatory regulation of the foreign trade turnover, of energy resources in the CIS includes: an agreement on the creation of resources and their effective use to ensure the parallel operation of the electric power systems of the CIS member states; Order of the Government of the Russian Federation "On signing the Protocol on the stages of the formation of the common electric power market of the CIS member states"; The Order of the Government of the Russian Federation "On the signing of the Agreement on the Organization, Management, Functioning and Development of the Common Market of Oil and Gas of the Member States of the Eurasian Economic Community"; Agreement "On general conditions for the supply of goods between organizations of CIS member states"; Agreement on the procedure for resolving disputes related to the implementation of economic activities; Convention of 22.01.1993 "On Legal Assistance and Legal Relations in Civil, Family Criminal Cases".

The studied documents of regulatory regulation, of foreign trade turnover of energy resources in the EEU cover: the Treaty on the Eurasian Economic Union; Federal Law No. 189-FZ of 11.07.2011 "On Ratification of the Agreement on the Organization, Management, Functioning and Development of Common Oil and Oil Products Markets of the Republic of Belarus, the Republic of Kazakhstan and the Russian Federation".

The choice of the applicable law to the cases of economic entities of the member countries of integration associations is disclosed in this chapter of the monograph on the basis of normative and doctrinal sources. The legal acts concerning regulated relations have been studied: Civil Code of the Kyrgyz Republic (Part II); The Civil Code of the Republic of Belarus; The Civil Code of the Republic of Moldova; Civil Code of the Republic of Uzbekistan; Civil Code of the Republic of Tajikistan; Civil Code of the Republic of Kazakhstan; Law of the Republic of Azerbaijan "On Private International Law"; Model Civil Code for the CIS member states (Part III); European Convention on International Commercial Arbitration. Also, scientific sources of a number of authors have been analyzed, among them: Aksenov A.G., Asoskov A.V., Vilkova N.G., Esenova G.Z., Zvekov V.P., Ivanov S.I., Iskakova Z. D., Kabasheva N.V., Marysheva N.I., Pavlova I.N., Salieva R.N., Yaroshenko K.B.

3 Methods

The scientific development of the content of this chapter of the monograph was carried out on the basis of the universal scientific method of historical materialism. Applied general scientific methods of cognition: dialectical, hypothetico-deductive method, generalization, induction and deduction, analysis and synthesis, empirical description, classification. The research also used private-science methods: legal-dogmatic, comparative legal analysis, etc.

4 Introduction

The external energy policy of the Russian Federation according to the Energy Strategy of Russia for the period until 2030 includes the development of cooperation in the field of energy with the countries of the Commonwealth of Independent States (hereinafter referred to as the CIS) and the Eurasian Economic Community (hereinafter—the EEU).

Since the formation of the Commonwealth, the formation of cooperation in the energy sector has always been given considerable attention. The CIS countries from the Soviet Union inherited cross-border transmission capacities and uniform technical standards. At the present stage, it is about creating an effective common market based on common principles. Legally, the CIS for 2018 includes 9 states and 2 states are in the status of countries that have not signed the treaty. This is an associated member of Turkmenistan and Ukraine, which has not signed the treaty. Perhaps, in 2018, Ukraine will formally also refuse to participate in the commonwealth, as Georgia refused in 2009.

In October 2014, the Eurasian Economic Community (EurAsEC) ceased to exist, the Eurasian Economic Union replaced it on January 1, 2015. At the level of the EEU there is a common market, the countries have a population of 183 million people. The union states—Russia, Kazakhstan and Belarus, as well as Armenia and

Kyrgyzstan—have mutually committed themselves to guarantee the free movement of goods and services, capital and labor, pursue a coordinated policy in the energy, industry, agriculture, transport.

5   Regulation of Foreign Trade Turnover of Energy Resources Within the CIS Countries

The energy sector sets the development of the economies of the CIS member states. The energy systems of the majority of the CIS participants operate in a parallel mode, electricity is transferred and exchanged with the energy systems of the neighboring countries of Europe and Asia, and also with China. Strengthening cooperation in the electric power industry facilitates the implementation of important interstate investment projects (Ivanov 2015).

Since February 1992, with the goal of harmonizing the external economic regulation of the energy sector, the CIS Electric Power Council acts as the main body coordinating interstate relations in the field of electric power industry. One of the tasks of the council is to assist the CIS participants in the unification and harmonization of regulatory legal acts in the electricity sector (Internet-portal of the CIS 2017).

In October 2002, the heads of government of the Commonwealth countries adopted the Agreement on Cooperation of the CIS Member States in the Field of Energy Efficiency and Energy Saving, in September 2004—Agreement on the creation of resources and their effective use to ensure a stable parallel operation of the electric power systems of the CIS member states (Bulletin of International Treaties, 2014. No. 5).

The decision of the Council of Heads of Government of the CIS of November 25, 2005 approved the Concept for the formation of a common electric power market of the CIS member states, which became an important instrument for the development of integration processes in the electric power industry. The concept, which takes into account the main principles of uniting the electricity markets in Europe, contains agreed approaches and views of the CIS countries to the formation of this market (Order of the Government of the Russian Federation No. 819-r of 20.05.2010).

With a view to implementing this Concept, an Agreement was concluded on the formation of a common electricity market of the CIS member states and an Agreement on the harmonization of customs procedures for the movement of electricity through the customs borders of the CIS member states. In practice, the conclusion of intergovernmental agreements with the CIS member states is widespread, through which the construction and operation of energy facilities is carried out.

On November 20, 2009, the CIS Council of Heads of Government approved the Concept of Cooperation of the CIS Member States in the Energy Sector (CIS Internet Portal), which defines the goals, principles, mechanisms and main directions of cooperation of the CIS member states in this field (with the exception of nuclear energy, the principles of cooperation in which are defined in the framework of the special program).

Priority of economic interests of the CIS member states in the energy sector is the main one in the Concept and is taken into account in the mechanisms of cooperation established by the Concept, such as:

- unification of norms, rules and technical regulations in the energy sector, including in the field of safety engineering;
- signing of bilateral and multilateral international treaties governing the relations of the CIS member states in the energy sector, as well as between the CIS member states and third countries;
- creation of the common energy market of the CIS member states, and first of all, the general electric power market;
- development and implementation of joint investment projects;
- exchange of experience and advanced technologies;
- establishment of agreed pricing rules for—energy resources and tariffs for their transportation;
- coordination of activities of the CIS member states in the energy sector within the CIS bodies, etc.

The events planned by the Concept confirm their consistency with the main directions of the medium and long-term outlook for the development of the energy regulation of the international integration association, identified in the previous chapter of the monograph, on the example of the EU—technological modernization of the fuel and energy complex, as well as the development of international cooperation.

In order to fulfill the Concept of Cooperation of the CIS Member States in the Energy Sector on November 20, 2009, the Council of Heads of Government of the CIS of May 21, 2010 approved the Plan of Priority Actions for its implementation (hereinafter—the Plan). The plan provides for the formation of a common electricity market for the CIS member states, the joint construction of power lines, the development and use of renewable energy resources, the development of harmonized technical regulations, the creation of the necessary conditions for attracting investments in energy and other important activities. The CIS Executive Committee is instructed to submit annually to the Economic Council of the CIS Information on the implementation of the Plan.

At present, the interaction in the energy sector is carried out within the framework of the Concept of Cooperation of the CIS Member States in the Energy Sector and the Plan of priority actions for its Implementation.

Due to the existing economic ties in the energy sector, the legal harmonization, unification and coordination of the legislations of the CIS countries are necessary measures not only in the sense of building a common legal space of the Commonwealth countries, but also for the purpose of convergence with the norms of international law. The basis of harmonization and unification of the legislation can be general principles of state regulation of relations in the sphere of primary energy resources, formulated in the framework of international approaches to legal regulation (Salieva 2013).

When regulating the relations arising from the foreign trade contract for the sale of goods between the economic entities of the CIS countries, two main methods of regulating relations that are the subject of private international law are used: the conflict method and the substantive law. The problem of choosing the applicable law to the contract within the framework of a substantive legal method is resolved by applying to the treaty international unified substantive rules that directly regulate the rights and obligations of the parties to the contract without seeking a conflict binding.

The problem of choosing the law applicable to the contract is not completely solved by the provisions of the Vienna Convention, the Agreement of 20.03.1992 "On general conditions for the supply of goods between organizations of member states of the Commonwealth of Independent States" (The Eutron Fund for Nomatological Documents 2017), as well as bilateral international agreements. The reason for this is the lack of unified norms regulating a number of fundamental issues. In addition, not all CIS member states are parties to these international agreements (Bulletin of International Treaties, 1993, No. 4.). The absence in the contract of a condition agreed by the parties on the applicable law leads to the need to resort to a conflict-based method of regulation (Aksenov 2012).

Priority in determining the applicable law to the foreign trade contract for the sale of goods between business entities of the CIS countries has the principle of autonomy of the will of the parties.

The UN Convention on Contracts for the International Sale of Goods and the 1992 Agreement establish the principle of autonomy of the will of the parties in a broad sense—as the principle of absolute freedom of contract, and not as an attachment formula.

This principle is reflected in the agreements concluded between the CIS countries, which contain provisions on the choice of the applicable law: the Kiev Agreement of 20.03.1992 "On the procedure for resolving disputes related to the implementation of economic activities" (clause "e" of Article 11) (Laws, Codes and Normative-Legal Acts of the Russian Federation 2017). Ten states of the CIS are parties to the Agreement: Azerbaijan, Armenia, Belarus, Kazakhstan, Kyrgyzstan, Russia, Tajikistan, Turkmenistan, Uzbekistan, Ukraine. (Informational Bulletin of the Council of Heads of State and the Council of Heads of Government of the CIS "Commonwealth". 1992. № 4).

Also, the principle of freedom of contract was reflected in the Minsk Convention of 22.01.1993 "On Legal Assistance and Legal Relations in Civil, Family and Criminal Cases" (Article 41). In the 1993 Minsk Convention, all CIS states participate, a number of changes to the Convention were made by the Protocol of 28.03.1997. It should be clarified that the Convention ceased to exist by virtue of paragraph 3 of Article 120 of the Convention on Legal Assistance and Legal Relations in Civil, Family and Criminal Cases of 07.10.2002. The new convention with the same name (adopted in 2002 in Chisinau—the Chisinau Convention of 2002) The Convention on Legal Assistance and Legal Relations in Civil, Family and Criminal Cases of 07.10.2002 entered into force for the states (as of June 24, 2016 of the year): Belarus, Azerbaijan and Kazakhstan—since April 27, 2004; Kyrgyzstan—since October 1, 2004; Armenia—since 19 February 2005; Tajikistan—since May 17, 2005 (Collected Legislation of the Russian Federation, 1995. № 17. Article 1472). In the Russian Federation, the Convention has been ratified by Federal Law No. 16-FZ of August 4, 1994.

The importance of the above-mentioned international treaties to regulate relations between the CIS countries can not be overemphasized. The Kiev agreement of 1992 covers issues related to the resolution of cases arising from contractual and other civil law relations between economic entities, subject to application of the law to contracts and provision of information on the law, issues of legal protection of economic entities, recognition and enforcement of decisions that have entered into legal force courts of one of the CIS member states, on the territory of other member countries of the Commonwealth.

The Minsk Convention of 1993 regulates the relations arising in the provision of legal assistance, the recognition and enforcement of foreign judgments in disputes involving legal entities related to the implementation of economic activities. It applies to any civil law disputes, unlike the Kiev Agreement of 1992, which regulates the resolution of disputes related to the implementation of economic activities.

According to Article 82 of the 1993 Minsk Convention (article 118, paragraph 3, of the Chisinau Convention, 2002), it does not affect the provisions of other international treaties to which the contracting parties are parties. Consequently, the courts of the CIS member states rely on the 1992 Kiev Agreement, as the main international act of a special action regulating, among other issues, the applicable law to the treaty under consideration. Since the Republic of Moldova is not a party to the Kiev Agreement of 1992, it applies the 1993 Minsk Convention when clarifying the applicable law.

The cooperation of the CIS member states in the field of energy should be based on the principles of equality and mutual benefit, as well as non-discrimination. Significant for cooperation is also compliance with market pricing principles. The CIS member states are taking action to create a common energy market for the CIS member states. Achieving this target is impossible without the unification of norms, the development of uniform rules and technical regulations, the harmonization of the pricing system on the basis of market principles. Priorities of cooperation of the CIS member states—technological development and strengthening of international relations with organizations and integration associations of the main external trade partner—the EU, include a number of activities. Creation and introduction of new high-tech power capacities, development of trans-national transport energy networks, expansion of cooperation in the development of oil products deposits, cooperation in the use of renewable energy resources, development of joint infrastructure projects, ensuring fuel and energy security and interaction in crisis situations in the energy sector, and improving energy efficiency.

However, the strengthening of the different-level and different-speed nature of economic integration in the post-Soviet space has become a factor in inhibiting the development of unified approaches to the implementation of energy policy in the CIS as a whole. In modern conditions, there is an intensification of cooperation in the energy sector of the countries of the Eurasian Economic Union, as a subregional integration association in the CIS space. The scale of the influence of this integration association on the energy sphere of the world economy is determined by the huge energy potential of the Union: it accounts for about 20% of the world's reserves and natural gas production and almost 25% of its exports, more than 20% of coal reserves and 6% of its production, 7% of the world's oil reserves, 14% of its production and 16% of exports, 5% of electricity production (Energy, 2015).

6 Regulation of Foreign Trade Turnover of Energy Resources in the Member Countries of the EEU

Another major international organization for regional economic integration with the participation of the Russian Federation is the Eurasian Economic Union (EEU).

The EEU is the legal successor of the Eurasian Economic Community (EurAsEC), established on its basis and entered into force on January 1, 2015. The purpose of the EEU is the formation of the Single Economic Space in the territory of the member states. This is stated explicitly in the Treaty on the Eurasian Economic Union, adopted in 2014 (the Treaty on the Eurasian Economic Union, 2014). At present, the EEU unites 5 states, including the Republic of Armenia, the Republic of Belarus, the Republic of Kazakhstan, Kyrgyzstan and the Russian Federation. Almost all the countries of the EEU are members of the WTO, and are obliged to comply with the norms, standards and recommendations of this organization. The only exception is the Republic of Belarus, which has not yet joined the WTO, but since 2016 has been actively negotiating accession to the WTO and, therefore, has a different order from other countries of the EEU in the legal regulation of foreign trade.

The volume of mutual trade of the countries of the Eurasian Economic Union in January–June of 2017 increased by 27% ($5.3 billion) as compared to January–June 2016. The most significant—1.6 times (by $1.2 billion)—increased mutual trade in metals and products from them. The mutual trade in machinery, equipment and vehicles increased by 36.5% ($1.1 billion), food and agricultural raw materials by 22.7% ($730 million), mineral products by 18% 6% ($1.1 billion). In general, the volume of mutual trade of the Eurasian Economic Union for January–June 2017 amounted to 25.1 billion dollars (Eurasian Economic Commission 2017).

According to Article 2 of the Treaty on the Eurasian Economic Union, "unification of legislation" is the approximation of the legislation of member states aimed at establishing identical mechanisms of legal regulation in certain areas defined by this Treaty.

The countries of the Eurasian Economic Community, including the Russian Federation, intend to move to a common oil and gas market and a single economic space by 2025, which will increase the volume of exports and imports of energy resources by several times. Thus, in the next decade, the EEU countries will become key partners of the Russian Federation in the sphere of trade in oil and gas resources. In this regard, the study of the sources of legal regulation of foreign trade in these countries becomes particularly relevant.

In the energy sector, the activities of the EAEC are aimed at the formation and joint development of a common energy market. Experts note that the uneven distribution of energy sources across the countries of the community determines the objective need for deepening integration in the fuel and energy sector of the economy (Iskakova et al. 2014). Its basis is the diversification of the energy complex, which ensures a steady growth of the economies of the countries of the community. The creation of a single energy space and the establishment of a free energy trade zone is one of the priority areas of cooperation among the EEU countries.

The basis of mutually beneficial energy cooperation between the EEU countries is the historically established close cooperation in the spheres of extraction, transportation and processing of fuel and energy minerals, the existence of a unified technological basis for the electric power industry and a developed electric grid infrastructure. This determines the main integration orientations and priorities of the EEU states in the energy sector: the implementation of a coordinated energy policy, and the creation of common energy markets, which include electricity, gas, oil and oil products markets. The formation of a common electricity market will be completed by July 1, 2019, and the creation of common markets for gas, oil and oil products is expected to take place by January 1, 2025 (Eurasian Economic Commission 2017).

The agreement on the Eurasian Economic Union was signed in Astana on May 29, 2014 and was amended from 08.05.2015, amendments and additions were made to it, which came into force on August 12, 2017. The agreement on the EEU became the most important document that constitutes the energy right of this union. In order to unify the legal regulation, uniform concepts are fixed, such as: "oil and oil products"; "Common market of oil and oil products of the Member States"; "Transportation of oil and oil products"; "Access to services of subjects of natural monopolies in the field of transportation of oil and oil products."

In order to harmonize the external economic regulation of the energy sector, Chapter 20 "Energy" was included in the Treaty establishing the EAEC, which establishes the basic principles of cooperation and trade in this field, which are detailed in Article 79. In accordance with Article 81 of the Treaty, the member states of the EEU carry out a step-by-step formation of the common electric power market, the result of which will be the signing of the relevant international treaty (the Eurasian Economic Union Treaty). The enactment of this normative legal act is projected for 2019. This event will become a new milestone in the development of the external economic cooperation of the EEU countries. We believe that the role of this agreement in the process of harmonizing the external economic regulation of the energy sector is enormous. In addition to the above aspects, it regulates the activities of natural monopolies in the electricity sector, the formation of gas and oil markets, as well as pricing in this sphere.

The provisions of the sections of the Treaty on general principles and rules of competition and on natural monopolies regulating relations in the field of electric power, gas, oil and oil products reflect a unified approach in these areas of regulation, are included in the system of norms and provisions forming the Union's energy law (Pavlova 2015).

The decision of the Higher Eurasian Economic Council of 08.05.2015 № 12 approved the Concept of the formation of a common electro-energy market, which determines the functional structure and stages of the formation of the common electric power market, its subject composition and system of regulatory acts, as well as the directions of interaction of the parties in the formation of this market (Eurasian Economic commission 2017).

The basic principles for creating the regulatory, organizational, technological and institutional base of the Union's common energy markets were approved by the Treaty on the Eurasian Economic Union of May 29, 2014. The formation of such markets implies ensuring the free movement of energy resources across the territories of the EEA states, creating a competitive environment in the energy sector, ensuring equal access to services of natural monopolies in the field of transportation and transit of energy carriers, joint development of the infrastructure of energy markets and the implementation of an agreed tariff policy.

The process of formation of common energy markets was initially divided into the following stages: first, the development and approval of an appropriate concept; then the development of a program for the implementation of the concept and the implementation of its activities; finally, the entry into force of an international treaty on the formation of such markets. The concept of the general electric power market of the Union was developed and approved on May 8, 2015. The Concept for the Formation of Common Oil and Oil Products Markets and the Concept for the Formation of a Common Gas Market was approved by the Supreme Eurasian Economic Council in Astana on June 2, 2016 at the level of the heads of the EEU member states.

The program for the formation of the common electric power market of the EEU was approved by the decision of the Higher Eurasian Economic Council on December 26, 2016 (Decision of the Supreme Eurasian Economic Council of December 26, 2016 No. 20).

The Council of the Eurasian Economic Commission approved the program for the formation of common markets for oil and petroleum products on December 22, 2017, to further submit it for consideration and approval to the heads of the member states of the association. The document contains over 50 events—from the harmonization of the legislation of the Member States in the oil sector to the development of software and hardware complexes of exchange traders and clearing organizations necessary for the implementation of exchange trades in oil and oil products. The next stage—from 2018 to 2023 years—will develop rules for access to oil and petroleum products transportation systems and rules of trade on common markets. The project activities should be implemented before 1 January 2024.

The program for the formation of the common gas market of the EEU was approved by the Consultative Committee on Oil and Gas of the Eurasian Economic Commission on April 24, 2017 (Eurasian Economic Commission 2017). This document takes into account the economic interests of the member states, including the features of the functioning and development of their gas markets, and the specific features of national legislation. The document includes a set of interrelated organizational, technological and other measures that ensure the formation of a common gas market of the Union, and contains a list of more than 30 events relating to the procedure for the implementation of exchange trading and rules for gas trading, the implementation of price and tariff policy in the general gas market, including the formation of exchange and OTC price indicators, as well as harmonization of the legislation of the Member States regulating the gas sphere.

Eurasian Economic Commission planned to complete the development of the program and its approval at the level of the presidents of the countries of the Union before January 1, 2018, respectively, to implement the program until January 1, 2024. However, according to the Commission officials for January 2018, "the discussion on the gas market formation program ... will be launched next year" (Eurasian Economic Commission 2017). The presence of controversial and unsettled issues in such a sensitive sphere of the economies of the EEU countries, as the gas sector, calls into question the possibility of observing the terms of the conclusion of the international agreement on the formation of the common gas market of the Union by the members of the EEU and its entry into force no later than January 1, 2025.

Nevertheless, the very idea of creating common markets for the energy resources of the ETU is not disputed, as this will make it possible to use the powerful energy potential of the Union more efficiently to solve the energy supply problems of the economies of the member states, expand export opportunities and transit potential, increase the sustainability of the energy sector and its infrastructure to external and internal impacts. This, in turn, will ensure increased competitiveness and, on this basis, the growing influence of the EEU countries in the global economy.

Formation of the general electric power market of the EAEC provides for the formation of Union legislation, including energy law.

According to the provisions of Section XII of the Concept for the Formation of the Common Electric Power Market of the Union, the system of acts regulating its general electric power market is: The Treaty on the Eurasian Economic Union, an international agreement on the formation of the Union's common electric power market, acts of the Union's apparatus that determine the functioning of the subjects of the Common Electric Power Market of the Union and are developed in accordance with the Program for the formation of the common electric power market of the Union.

Judging by the subject of agreements concluded in the oil and gas sector between the countries at the present stage, it can be concluded that the direction of mutual cooperation has changed in the conditions of countries' striving for the organization and development of common markets.

Thus, in September 2005, an Agreement was signed on the organization, management, operation and development of a common oil and gas market between the member states of the EEU, which provides for the creation of a common market for oil and gas on the basis of national markets, production and resource potentials of fuel and energy complexes of the participating countries, transport communications of participants for the transit of oil and gas resources (Collection of Legislation of the Russian Federation, 2005. No. 40. Article 4082.).

This market is aimed at the free movement of gas, oil and products of their processing in the territories of the EEU member states, which is based on the application of the regime without tariff and non-tariff restrictions in trade in resources, information support systems, equal conditions for access to the infrastructure of national energy markets.

At present, more than 50 bilateral agreements have been concluded in the field of mutual trade in gas, oil and petroleum products on issues of supply, transportation (including transit), export of energy resources, formation of prices, as well as the procedure for payment and transfer of export customs duties, which, among other things, regulate relations in the field of export (re-export) of oil and oil products from the territories of member states outside the customs territory of the Union, establish restrictions (quotas) on the export/import of certain groups of goods in the mutual trade of member states (Pavlova 2015).

The legal basis for the common oil and gas market of the EEU member states was the Decision of the EurAsEC Interstate Council of 09.12.2010 No. 65 "On the implementation of the Action Plan for the formation of the Common Economic Space of the Republic of Belarus, the Republic of Kazakhstan and the Russian Federation". The document approves the list of agreements forming the Common Economic Space of the Republic of Belarus, the Republic of Kazakhstan and the Russian Federation and defining, among other things, the fundamental legal principles of their obligations in the energy sector, which presuppose a uniform civil law regulation, unified by the legislation of the member countries at the national level. In this list, the Agreement on the creation of conditions in the financial markets for free flow of capital, the Agreement on Harmonized Principles of Monetary Policy, the Agreement on Unified Principles and Rules of Competition, the Agreement on Uniform Principles and Rules for the Regulation of the Activity of Natural Monopoly Entities, Agreement on the rules for access to natural monopolies in the field of gas transportation through gas transmission systems, the Agreement on the organization, management, functioning and development of common oil and oil products markets of the Republic of Belarus, the Republic of Kazakhstan and the Russian Federation, and others.

In July 2011, Russia ratified the Agreement on the Organization, Management, Functioning and Development of Common Oil and Oil Products Markets of the Republic of Belarus, the Republic of Kazakhstan and the Russian Federation (Collection of Legislation of the Russian Federation, 2011). The basic principles of the Agreement are expressed in the non-use by the parties in mutual trade of quantitative restrictions and export customs duties, priority provision of the needs of the countries of the Common Economic Space in oil and oil products, unification of norms and standards for these goods, and guarantees of the possibility of long-term transportation of oil and oil products.

One of the main forms of implementing measures to ensure that States have undertaken their obligations are international agreements. International agreements are designed to strengthen interstate integration. Russia's ratification among the first agreements on the organization, management, functioning and development of common oil and oil products markets of the Republic of Belarus, the Republic of Kazakhstan and the Russian Federation confirms the great importance of these markets for the security and economic stability of the parties to the Agreement.

7 Peculiarities of Conflict Regulation of Relations Between Economic Entities of the CIS Countries and the EEU

The development of foreign economic relations between the CIS countries and the EEU in the energy sector opens serious prospects for Russia as the largest exporter of energy. The creation of the Eurasian Economic Union marked the beginning of a new stage of Russia's cooperation within the framework of the regional integration association. Just like the CIS, the EEU is an international organization that has united the countries of the post-Soviet space that have similar legislation and a vision of international politics.

In this regard, it will be useful to address the national legislative approaches of the countries participating in both integration associations.

Conflicting regulation of the contract of international sale of goods between economic entities is in the national legislation of the CIS countries, namely in civil codes and laws on private international law. These legislative acts contain rules concerning the autonomy of the will of the parties to the treaty. The main rules are:

1. The parties to the treaty are free to choose the law to be applied to the contract. Such a choice of the party of the contract can make both at the conclusion of the contract, and in the subsequent. Parties are also given the opportunity to change the applicable law.
2. The choice must be expressly expressed or definitely derived from the provisions of the treaty or from the circumstances of the case.
3. The parties to the contract may specify the applicable law both for the treaty as a whole and for its individual parts.

In consideration of the priority of the choice of the law applicable to the contract by agreement of the parties, paragraph 1 of Article 1198 of the Civil Code of the Kyrgyz Republic (Civil Code of the Kyrgyz Republic of 5 January 1998 No. 1 (PartII), Paragraph 2017a), paragraph 1 of Article 1124 of the Civil Code of the Republic of Belarus (Laws of the Republic of Belarus 2017), Part 1 of Article 1189 of the Civil Code of the Republic of Uzbekistan (Civil Code of the Republic of Uzbekistan, Lex 2017), paragraph 1 of Article 1218 of the Civil Code of the

Republic of Tajikistan (Civil Code of the Republic of Tajikistan of March 1, 2005, Paragraph 2017c), paragraph 1 of Article 1112 of the Civil Code of the Republic of Kazakhstan (Paragraph 2017d), established by law.

The Civil Codes of Armenia (Articles 1258, 1259), Belarus (Articles 1099, 1100), Kazakhstan (Articles 1090, 1091), Kyrgyzstan (Articles 1173, 1174), Moldova (Articles 1581, 1582), the Russian Federation (Articles 1192, 1193 of the Civil Code (Articles 1164, 1165), Tajikistan (Articles 1197, 1198), Law on Private International Law of Azerbaijan (Articles 4, 5) (Law of the Republic of Azerbaijan "On Private International Law" of 2000, The law of the Higher School of Economics 2017) contain prescriptions that restrict the validity of certain rules of foreign law. When the consequences of their application clearly contradict the public order of the state (reservation of a public order), as well as due to the mandatory application of mandatory norms of the country's legislation when specifying this in the mandatory norms themselves or in view of their special importance, including for ensuring the rights and legally protected interests of participants in civil traffic.

Since the provisions of the Model Civil Code for CIS member states were largely used in the development of the above-mentioned regulatory legal acts, the rules governing the application of the right to choose parties to the agreement are largely similar or slightly different from each other (Model Civil Code for the CIS Member States (Part Three).

Parties to the contract are not required to choose the law applicable to the contract. In the absence of agreement of the parties on the applicable law to the contract subject to application to the contract of sale of goods between economic entities of the Commonwealth countries, the law is determined on the basis of conflict rules of international treaties to which CIS states are participants (the Kiev Agreement of 1992, the Minsk Convention of 1993, Chisinau Convention 2002).

These international acts include conflict-of-laws rules for determining the rights and obligations of parties to a transaction under the law of the place of transaction, unless otherwise established by agreement of the parties.

The same binding was found in legislative provision in Article 126 of the Fundamentals of Civil Legislation of the USSR in 1961. In most countries, the use of this ineffective collision linkage has been abandoned, since it does not correspond to the current trends in the development of conflict of laws law, reflected in universal and regional conventions and individual conflicting interests of the parties (Vilkova 1997). At the present stage of the conclusion of contracts, often between absent parties by means of sending an offer and an acceptance, contracts lose contact with a particular place, and it is difficult to determine the location of the transaction.

In the competent scientific community it is noted that in determining the applicable law (Marysheva and Yaroshenko 2014), international commercial arbitration by virtue of paragraph 1 of Article VII of the European Convention on Foreign Trade Arbitration of 1961, of which the CIS countries are participants Azerbaijan, Belarus, Kazakhstan, Moldova, the Russian Federation and earlier Ukraine, independently determines the relevant conflict of laws (the European Convention on Foreign Trade Arbitration). In this connection, the priority is not the

1992 Kiev Agreement and the 1993 Minsk Convention, but the European Convention on Foreign Trade Arbitration, which is lex specialis for international commercial arbitration. It should be clarified that on November 9, 2016 a bill on Ukraine's withdrawal from the CIS introduced in the Verkhovna Rada of Ukraine.

Since the Kiev Agreement of 1992, the Minsk Convention of 1993 and the 2002 Chisinau Convention do not cover the regulation of such matters of a binding statute as interpretation, execution of the contract, the consequences of failure to perform or improper performance, termination of the contract and the consequences of its recognition as invalid, there is a need to appeal to the national conflict rules of the CIS countries.

Thus, in determining the law applicable to the treaty, in the absence of an agreement of the parties, the Law of the Republic of Azerbaijan "On Private International Law" (Article 25), the Civil Code of the Republic of Armenia (Article 1285), the Civil Code of the Republic of Belarus (Article 1125), the Civil Code of the Kyrgyz Republic 1199), the Civil Code of the Republic of Kazakhstan (Article 1113), the Civil Code of the Republic of Uzbekistan (Article 1190) and the Civil Code of the Republic of Tajikistan (Article 1219) similar to Article 1225 of Part III. The Model of the Civil Code for the CIS countries refer to the law of the country where the seller is the domicile or principal place of business.

Article 1211 of the Civil Code contains other criteria for determining the applicable law to contractual obligations in the absence of agreement between the parties. So, according to paragraph 1 of Article 1211 of the Civil Code of the Russian Federation, unless otherwise provided by this Code or another law, in the absence of an agreement of the parties on the law to be applied to the treaty, where, at the time of the conclusion of the contract, the place of residence or principal place of business of the party performing the performance that is of decisive importance for the content of the contract is located. Such party to the contract of sale is the party that acts as the seller.

The combination of a "flexible" conflict of laws rule that refers in the form of a general rule to the law of the country with which the treaty is most closely linked to "formalized" bindings that differentiate the types of contracts by the terms "parties whose execution is crucial for the content of the contract" allows to solve conflict questions, taking into account the peculiarities of each specific situation, and to limit the limits of judicial discretion at the same time (Zvekov and Marysheva 2002; Aksenov 2010).

In the absence of agreement of the parties on the choice of the applicable law, the conflict problem on contractual obligations is solved by means of the combination of the theory of characteristic execution and the principle of the closest relationship known long ago in international private law. In the Civil Code of the Russian Federation, these principles are given more flexible, taking into account the diversity of situations (Zvekov and Marysheva 2006).

Consequently, conflicts of laws of the national legislation of the majority of CIS member states that regulate the application of law to the contract of international sale of goods between economic entities of the CIS countries contain a uniform rule of their application leading to the law of the seller country. The exception is Article

1611 of the Civil Code of the Republic of Moldova, which, in the absence of an agreement between the parties, determines the general rule on the application of the law of the closest relationship to contractual obligations (Civil Code of the Republic of Moldova, 2002, Paragraph 2017b). At the same time, under the law of the most intimate connection is understood the law of the state in which the debtor at the time of the conclusion of the contract had a residence or location or was registered as a legal entity.

Contained in the Kiev Agreement of 1992 and the Minsk Convention of 1993 (the Chisinau Convention of 2002), the conflict binding does not correspond to the current trends of private international law. The use of an outdated binding to the place of conclusion of a treaty in modern conditions looks like an unjustified anachronism, which prevents the development of more progressive national conflict regulation, since in the event of a dispute resolution from a treaty involving parties from CIS member states, national courts are forced to apply outdated norms of international treaties, rather than more effective national conflict rules (Asoskov 2011). Differences in conflict binding in comparison with the norms of the European Convention of 1961 and the national legislation of the CIS countries create legal uncertainty in the regulation of the contract of international sale of goods between business entities of CIS countries.

The conflict-legal method of regulation as a whole does not allow to solve the problem of overcoming differences in the regulation of relations when concluding contracts between economic entities of different CIS countries, because even when the norms of national legislation are brought together, the legal system of each state retains certain features and differences.

A single geopolitical and geostrategic space, a common historical past, general legal and economic interests of economic entities of the CIS member countries, the nature of economic ties, trade turnover, strongly dictate the need to create a unified unified material and legal regulation (Asoskov 2011).

8 Conclusion

Thus, within the framework of the deepening of integration in the Eurasian space and the implementation of new directions for the interaction of member states, the improvement and development of the Union's energy law will be required, including with regard to the implementation of a coherent energy policy, ensuring non-discriminatory access to the common markets of the Union's energy resources by unified rules for the economic entities of the Union, developing and implementing unified rules for trade in energy resources, including exchange trade, infrastructure development and implementation of large joint cross-border projects on transport, mining, and related energy services (Pavlova 2015).

In conclusion, it should be emphasized that for Russia, as one of the world's largest energy exporters, the development of foreign economic relations with the CIS countries and the EEU has an unconditional priority and great prospects.

Integration and creation of an effective common energy market are impossible without the formation of a common energy policy of the member states of regional associations. Integration and common energy policy are the basis for the international unification of norms, harmonization of national laws and standards in the energy sector.

It is obvious that the development of external economic relations between the CIS countries and the EEU in the energy sector opens up serious prospects for Russia as the largest exporter of energy. The CIS and the EEU are international organizations that have united the countries of the post-Soviet space with a similar vision of international politics and economic development, including in the energy sector. The analysis of the legal regulation of the foreign trade turnover of energy resources of member countries of regional integration associations, based on the example of the Republic of Armenia and the Republic of Kazakhstan, confirmed the use of legal mechanisms similar to Russian law, including the structure of civil legislation, as well as approaches to contractual regulation and applicable law issues, uniformity in the post-Soviet legal space.

All recommendations, including priority directions and measures for their implementation, concerning the development of legal regulation of the foreign trade turnover of the energy resources of the Russian Federation are applicable to the legal regulation of relations between the CIS member countries and the EEU. At the same time, it is necessary to take into account the more developed, in comparison with the CIS, character of the international integration of the countries that make up the EEU. It is confirmed by their readiness to create a general electricity market by July 2019, and by 2025—the common market of gas, oil and oil products. In this regard, in the process of creating a common energy regulation of the CIS countries, emphasis should be placed on its harmonization through the development of program framework acts. Regarding the EEU, a more effective method is the unification of norms through the conclusion of bilateral and multilateral agreements and the development of legally binding standards.

References

Agreement on the creation of resources and their effective use to ensure the parallel operation of the electricity systems of the CIS member states 2014. *Bulletin of International Treaties,* no. 5, Art. 1314.

Aksenov, A. G. (2010). *Treaty on the international sale of goods between entrepreneurs of the countries of the commonwealth of independent states.* Thesis, Moscow.

Aksenov, A. G. (2012). *Agreement on the international sale of goods between business entities of the CIS countries.* Moscow: Publishing House "Infotropic Media".

Asoskov, A. V. (2011). *Normoobrazuyuschie factors affecting the content of conflict regulation of contractual obligations.* Thesis, Moscow.

CIS portal. (2015). *Ivanov SI theses of the speech of the deputy chairman of the CIS executive committee at the Asia-Pacific energy forum on May 27–30, 2013*. Retrieved November 25, 2015, from http://www.unescap.org/sites/default/files/4.%20Sergey%20Ivanov%20(RUS). PDF.

Convention. (1993). *On legal assistance and legal relations in civil, family criminal cases, (ratified by Federal Law of August 4 1994, no. 16-FZ).*

Eurasian Economic Commission. (2017). *Treaty on the Eurasian Economic Union [in red. from 08/05/2015].* Retrieved September 01, 2017, from http://www.eurasiancommission.org.

European Convention on Foreign Trade Arbitration (Geneva, April 21, 1961). (1964). *Gazette of the Supreme Soviet of the USSR*, no. 44, Art. 485.

Federal Law. (2011). *On ratification of the agreement on the organization, management, functioning and development of common oil and oil products markets of the republic of Belarus, the Republic of Kazakhstan and the Russian Federation (adopted by the State Duma on July 1, 2011, no. 189-FL).*

Internet-portal of the CIS. (2017). http://www.e-cis.info/.

Iskakova, Z. D., Kabasheva, N. V., & Esenova, G. Z. (2014). On integration association of the EAEC participants: Expectations and implementation mechanisms. *International Journal of Applied and Fundamental Research, 10–3*, 19–21.

Laws of the Republic of Belarus. (2017). *Civil code of the republic of Belarus*. Retrieved February 25, 2017, from http://bel-kodeksy.com/gk_rb.htm.

Laws, Codes and Normative-Legal Acts of the Russian Federation. (2017). *Agreement of 20.03.1992 "On the procedure for the resolution of disputes related to the implementation of economic activity".* Retrieved November 25, 2017, from http://legalacts.ru/doc/soglashenie-stran-sng-ot-20031992-o-porjadke.

Lex. (2017). *Civil code of the republic of Uzbekistan*. Retrieved February 25, 2017, from http://www.lex.uz/mobileact/188050.

Marysheva, N. I., & Yaroshenko, K. B. (2014). *Commentary on the civil code of the Russian federation, Part three (itemized)*. Moscow: Law Firm CONTRACT.

Model Civil Code for the CIS Member States. (Part Three). (adopted by the Resolution of the Interparliamentary Assembly of the CIS Member Nations 1996, *Annex to the Information Bulletin of the IPA of the CIS Member States*, 10, 3–84.

Order of the Government of the Russian Federation. (2010). *On signing the Protocol on the stages of the formation of the common electricity market of the CIS member states (approved by the Order of the Government of the Russian Federation of May 20, 2010, no. 819-r).*

Paragraph. (2017a). *Civil code of the Kyrgyz Republic of January 5, 1998, no. 1 (Part II) (as amended on March 9, 2017).* Retrieved February 25, 2017, from http://online.adviser.kg/ Document/?link_id=1000864630.

Paragraph. (2017b). *Civil code of the republic of Moldova of June 6, 2002, no. 1107-XV.* Retrieved March 25, 2017, from http://online.zakon.kz/Document/?doc_id=30397921.

Paragraph. (2017c). *Civil code of the republic of Tajikistan of 1 March 2005.* Retrieved February 25, 2017, from http://online.zakon.kz/Document/?doc_id=30447965#pos=40;-159.

Paragraph. (2017d). *Civil code of the republic of Kazakhstan*. Retrieved February 25, 2017, from http://online.zakon.kz/m/document/?doc_id=1013880#sub_id=11120000.

Pavlova, I. N. (2015). Basic aspects of the formation of the energy law of the Eurasian Economic Union. *Energy Law, 2*, 45–48.

Salieva, R. N. (2013). Legal and environmental aspects of regulation in the sphere of primary energy sources in the framework of the energy strategy of Russia. *Lawyer, 21*, 27–31.

The Eutron Fund for Nomatological Documents. (2017). *The Agreement of 20.03.1992 "On general conditions for the supply of goods between organizations of the member nations of the commonwealth of independent states".* Retrieved November 25, 2017, from http://docs.cntd.ru/ document/1900461.

The law of the Higher School of Economics. (2017). *The law of the republic of Azerbaijan "On private international law" of 2000*. Retrieved February 25, 2017, from https://pravo.hse.ru/intprilaw/doc/070801.

Vilkova, N. G. (1997). Unification of conflict rules within the CIS. *Journal of Russian Law, 10,* 91–97.

Zvekov, V. P., & Marysheva, N. I. (2002). New codification of norms of private international law. *Economics and Law, 6,* 3–16.

Zvekov, V. P., & Marysheva, N. I. (2006). Law of Ukraine 2005 on "International Private Law". *Economics and Law, 5,* 129–138.

General Energy Policy and Ways of Development of Legal Regulation of Foreign Trade Turnover of Energy Resources of the BRICS Countries

Agnessa O. Inshakova and Igor P. Marchukov

1 Annotation

The chapter is devoted to the study of civil-law regulation of the foreign trade turnover of energy resources of the BRICS countries, which is currently becoming one of the priority areas for strengthening and developing foreign economic relations of Russia and is a powerful and dynamically developing alliance that has an impact on energy markets. The analysis of the unified legal support for the foreign trade turnover of the energy resources of the BRICS countries analyzed. The absence of a unified international regulatory framework based on adopted general policy documents is stated. The process of formation of unified legal mechanisms aimed at regulating the foreign trade turnover of the energy resources of member countries based on a common energy policy studied. The study showed, that the agreements operating within the framework of the BRICS integration association are, in fact, program acts, that establish only general directions for the development of cooperation between the participating countries, and do not contain specific provisions for civil-law regulation of foreign trade relations in the energy sector.

It is noted, that the difficulties associated with the formation of a unified legal space for the foreign trade turnover of the energy resources of the BRICS countries at the present stage are a consequence of the lack of the legal personality of the new type of integration structure. Other factors, that hamper the acceleration of the processes of unification of national legislation in the sphere of civil and international private law,

A. O. Inshakova (✉)
Department of Civil and International Private Law, Institute of Law,
Volgograd State University, Volgograd, Russia
e-mail: gimchp@volsu.ru; ainshakova@list.ru

I. P. Marchukov
Department of Civil and International Private Law, Volgograd State University, Volgograd,
Russia
e-mail: gimchp@volsu.ru

© Springer International Publishing AG, part of Springer Nature 2019
O. V. Inshakov et al. (eds.), *Energy Sector: A Systemic Analysis of Economy,
Foreign Trade and Legal Regulations*, Lecture Notes in Networks and Systems 44,
https://doi.org/10.1007/978-3-319-90966-0_15

also identified, which, among other things, determine the differences in legislative approaches and the choice by the national legislator of methods and means for the legal regulation of foreign trade in energy resources.

The degree of unification of the civil and legal regulation of the foreign trade turnover of the energy resources of the BRICS countries is studied on the basis of the Vienna Convention on Contracts for the International Sale of Goods of 1980. National legislative, including collision, bases of the foreign trade regulation of the countries-participants BRICS are investigated.

It is concluded, that civil-law regulation of purchase and sale in the BRICS countries can hardly be called unified. However, the fact of participation in the Vienna Convention of most of the countries of the association and the possibility of its application in these countries, the uniformity in national legislative approaches regarding the recognition of treaties (deals, agreements) related to the foreign trade turnover of energy resources that define them as prisoners, as well as the grounds for their termination, create the necessary basic principle of a unified foreign trade turnover of energy resources. Based on the results of a comparative legal analysis of Russian civil legislation and civil legislation of the BRICS countries in the part of unified rules of international trade, first of all, the issues of applicable law to contractual obligations of the parties (including in the absence of the choice of the right by the parties and the establishment of a mandatory form of an international sales contract, as well as the difficulties encountered in this regard in regulating the obligations of the parties) concludes that the unification of the legal regulation of foreign trade relations of the BRICS countries in the field of energy at this stage should be concentrated in the national law of the participating countries.

2 Materials

Due to the empirical nature of the content of this chapter, the materials that formed the basis for its development are basically legal instruments of a strategic, programmatic and regulatory nature of international and domestic regulatory levels. Among the international legal instruments used in the study: The Delhi Declaration of the 4th BRICS Summit adopted on the results of the Summit on December 28–29, 2012, which consolidated the foundations of energy security in the BRICS countries, the BRICS Economic Partnership Strategy (Ufa's Declaration of the VII BRICS Summit), Joint Statement of the Russian Federation and the People's Republic of China "On Mutually Beneficial Cooperation and Deepening the Relationship of Comprehensive Partnership and Strategic Cooperation".

The intrastate regulatory level represented by the Concept of the Russian Federation's participation in the BRICS unification, as well as by the normative acts containing the rules of national conflict regulation in the sphere of the obligatory energy relations, among which: The laws of the People's Republic of China "On foreign trade" in 1994, the main law of South Africa, regulating foreign economic activity "Act on the Administration of Foreign Economic Activities", 2002, the main law of India on foreign economic activity. The Act "On the Development and Regulation of Foreign Trade" 1992.

In addition, analytical and statistical data published in the Energy Bulletin, as well as on official websites of the Government, Ministries and departments of the Russian Federation contributed to the achievement of the research results. Thus, the data of the Analytical Center under the Government of the Russian Federation, the Ministry of Energy of the Russian Federation, the Official Site of RIA Novosti, the Official Site of the Chairmanship of the Russian Federation in the BRICS used.

The theoretical basis of the research was the works devoted to the actual problems of the legal regulation of the current energy relations of the Russian Federation as a whole, including those with an international character, including works by Vysotsky V., Gudkov I.V., Lakhno P.G., Sevastyanova T.L., Plakitkin V., Ujanaev S.V.

Geopolitical aspects, difficulties and prospects for legal regulation in the energy sphere directly from the BRICS countries were studied thanks to publications of scientists, including: I.P. Marchukov, M.O. Ryazanova, D.D. Startsev, S.V. Sharko.

Problems of conflict regulation, issues of applicable law, as well as general provisions of civil law regulation of the international contract for the sale and purchase of energy resources of the BRICS countries were studied by reference to the provisions of the Vienna Convention of 1980 and to the scientific works of Agusto KLL, Belikova KM, Costa Lazota LA, Rosenberg MG. and etc.

3 Methods

The work uses the universal scientific method of historical materialism. To the general scientific methods of the research carried out include methods: dialectical, hypothetico-deductive, generalization, induction, deduction, analysis, synthesis and empirical description.

Among the particular scientific methods applied in the process of scientific development of this chapter: economic and statistical, the method of evaluating legal processes, the formal legal, interpretational, etc. The main, used in the chapter arsenal of private scientific methods, is the method of comparative legal analysis. This method was used, first of all, in the process of comparing the provisions of international legal treaties, recommendatory acts of non-normative program nature and domestic sources of the BRICS countries. With the aim of obtaining conclusions about their coherence, the presence of collisions and the implementation at the national level of the basic principles defined by the countries of the common energy policy of the integration association. In addition, in the process of comparative legal analysis, the rules for the regulation of international contracts for the sale and purchase of BRICS countries were compared, as well as related to their conclusion and implementation of conflict-of-laws regulations and issues applicable to the obligations of the parties to the law.

4 Introduction

BRICS[1] (*earlier—BRIC*) is a powerful alliance of dynamically developing countries that have embarked on the new industrial standards of countries demonstrating high economic growth and rapidly growing influence in regional and global markets, including energy markets. BRICS includes five countries: Russia, Brazil, India, China and South Africa (South Africa). Maintaining and developing energy cooperation for Russia and the European Union plays an important role in resolving its energy problems, strengthening its competitive positions in the global energy sector and developing the national fuel and energy complex. At the same time, the interstate integration association with the participation of the Russian Federation BRICS plays a key role in the development of the entire world fuel and energy complex, primarily because of the export opportunities of Russia and Brazil. For Russia, cooperation within the BRICS is of particular interest in the energy sector, the implementation of which primarily depends on the competently constructed, including legal methods and means of the common energy policy, as well as the exchange of experience and technologies of the energy industry (Sharko 2011; Carr 2013). In addition, we should not forget about the aggravated state of the current geopolitical situation, characterized by the growth of both contradictions between the Russian Federation and the EU, as well as cross-country contradictions within the framework of a pan-European integration space. Which change priorities and make it necessary to more closely develop the external economic relations of the Russian Federation in the energy sector under the aegis of the new international BRICS union on an equal footing (Leal-Arcas et al. 2015).

BRICS is becoming one of the preferred areas for strengthening and developing foreign economic relations of Russia, which form the potential for the development of intercountry cooperation in the trade and production spheres. To date, it is an international association unique in terms of scale, comparable in its authority to international associations of global scale, such as the G20 and G8 (Gudkov and Lakhno 2008), second only to the Group of Seven and the commonwealth of developed OECD countries. BRICS is a kind of mutually beneficial option of legal, political and trade cooperation of states. It is dictated by the prevailing international situation, as well as the time required by compromise (Plakitkin 2011).

Undoubtedly, BRICS has a huge potential in international energy relations. It is based on voluminous reserves of energy and natural resources that can not be perceived otherwise as a significant reserve for prospective and multifunctional cooperation. This potential provides a material and technical basis for entry into a post-industrial society, which is built on the priorities of innovation based on multi-industry production.

---

[1]The BRICS is an abbreviation consisting of the capital letters of the international organizations of the countries—Brazil, Russia, India, China and the Republic of South Africa. Until 2011, marked by joining the regional international organization of South Africa, in reference to its designation, the abbreviation BRIC (Note by the authors) was used.

5 Conditionality and Priority Measures for the Development of the Common Energy Market of the BRICS Countries

Currently, the BRICS countries provide about 35% of the world's energy production and consumption, and according to the estimates of the "Forecast of the development of the world's energy and Russia up to 2040" by ISER RAS and the Analytical Center, by 2040 this share will make up almost 45% (Analytical Center under the Government of the Russian Federation 2017a).

It was the power industry, according to most experts, which was to form the economic basis for practical integration. At the stage of the initial formation of the BRICS block, the energy sphere was perceived as the main source of mutual benefits. Thus, the integration association was predicted by the future of a certain "energy club". But, despite the fact that BRICS has always been recognized as a potentially powerful player in the world energy markets, it has never been considered only as an energetically oriented association. Creating conditions for highly effective cooperation and strengthening the technological and economic potential of participating countries that will ensure sustainable economic development and strengthen financial and social stability within countries through mutual fair economic integration and sectoral coordination is the main target of BRICS (Inshakova et al. 2015).

As a key for Russia, the energy export industry must meet the requirements of the 21st century, oriented to a qualitatively renewed fuel and energy complex of the country—cost-effective and innovative, equipped with advanced technologies and highly qualified personnel. The most important direction of energy policy since the end of the XX century are the problems of increasing energy efficiency and energy saving. In turn, in order to coordinate the long-term stable provision of the economy of the country and the population with all types of energy, its internal and external economic turnover, it is necessary to have a long-term energy policy, scientifically substantiated, accepted by society and state power institutions (Lakhno 2012; Marchukov 2015).

The interest of Russia in the development of the energy market within the BRICS framework can be explained by a number of factors.

First, Russia is one of the largest producers and exporters of oil, natural gas, coal to the world market, which is interested in ensuring the stability of supplies, including in their expansion, as well as in reducing the volatility of world commodity prices.

Thus, in April 2016, oil production and exports increased by 1.4% and 10.8%, respectively, compared to April 2015. The export of natural gas continued to grow —in March 2016 its volumes increased by 7.0% compared to March 2015, and by the results of three months the growth was 13.4% compared to January–March 2015 (Analytical Center under the Government of the Russian Federation 2017b). In general, in 2016, according to the Ministry of Energy of Russia, oil production in Russia amounted to 547.3 million tons, which is 2.6% more than the value of 2015 and is the maximum indicator for the post-Soviet period (Analytical Center under the Government of the Russian Federation 2017c).

As for the far abroad countries, in 2016 Russia increased its oil export by 6.6% in annual terms to 236 million 195.8 thousand tons, while natural gas exports increased by 13.8% to 164.7 billion cubic meters (News. The economy 2017). In general, in 2016, Russia supplied abroad 254 million 767.4 thousand tons of oil for $74 billion and 198.7 billion cubic meters of gas by $31.3 billion (News. The economy 2017).

Secondly, the intensive development of cooperation with the BRICS countries in the energy sector was influenced by the events of 2014, marked by the imposition of sanctions against Russian oil and gas companies, which entailed the risk of underinvestment in the energy sector. In addition, there was an acute need to search for new markets for oil and gas products, as well as sources of technology and foreign investment in the energy sector. In this regard, there has been a strong demand for energy in rapidly developing countries such as China and India (Vysotsky 2012).

Realizing cooperation in the main areas with the BRICS countries in the sphere of foreign trade in energy resources, Russia is undertaking a whole range of priority measures. First of all, this is the development of oil and gas projects in new areas of resource development (the Arctic shelf, the Far East), which require investment in the Russian fuel and energy complex. Not the last place in the complex of intercountry cooperation is the sharing of experience in the field of gas and oil production, technology import. In addition, gas supplies to the BRICS countries are gradually increasing, primarily to China and India. And, finally, serious attention is paid to the development of the global functioning of not only global, but also regional energy markets, as well as the implementation of mechanisms for the absence of sharp fluctuations in these markets, primarily in the price policy for raw materials.

President of the Russian Federation V.V. Putin at a meeting of the BRICS in an expanded format on October 16, 2016 announced the need to work out an initiative to create an energy agency BRICS (RIA official website 2016).

Thus, the BRICS countries show increased interest in the development of international cooperation in the field of energy conservation, as well as energy-efficient technologies and increasing energy efficiency. In the energy sector, the BRICS countries have plans to work together according to national priorities to facilitate the use of renewable energy by sharing experience in the use of renewable energy and international cooperation, cooperation in the field of professional training, technology transfer in the energy sector and provision of advisory services, researches and developments (Marchukov 2015).

The development of foreign trade turnover of energy resources in the BRICS countries is based not only on activities of a narrow profile. Successful energy cooperation between the united countries and the interests of Russia is facilitated by factors and measures of a general political nature aimed at strengthening state security and economic power in general, as well as contributing to the development of all branches of the national economy without exception.

Indeed, the role of the BRICS for Russia is due to the prospects of integrating the Russian economy into the Asian markets; creation of a base for the formation of available reserves for capital investments in the internal economy of the country; priorities of foreign exchange and monetary policy; common interests with states that can influence decision-making in international structures, and are palpable for the world economy.

In the process of developing recommendations on the unification of the legal regulation of the foreign trade turnover of the BRICS countries, it should be borne in mind that the association has property and legal independence, does not have a formal definition, which makes the BRICS policy free from deterrence, and the assurances and guarantees they have accepted can be violated without applying counter sanctions. Also, the creation of international institutions (a bank, a reserve fund) with the participation of the BRICS countries promotes the development of foreign economic relations, interaction in common interests and strengthening of the banking systems of states (Marchukov 2015).

All of the above gives grounds to believe that the consolidation of BRICS contributes to the strengthening of Russia's global economic positions and can become one of the most powerful tools for the development of its foreign trade turnover, primarily in the energy sector (Sevastyanova 2013).

Unfortunately, at the moment from among the BRICS countries only China is a major foreign economic and, in particular, foreign trade partner of Russia. Thus, in the countries of the association, about 10% is the total exports from Russia and 20% of imports. In this case, without China, these figures would be less than 2 and 3% respectively. The existence of such a low level of mutual trade is due, first of all, to territorial remoteness and the absence of established historical ties of the countries participating in the informal international association (Uyanaev 2013).

Returning to the characteristics and forecasting of the prospects of BRICS as an effective platform for developing cooperation in the foreign trade turnover of energy, it will be useful to recall that the international division of labor takes place between three producers (Brazil, Russia, South Africa) and two energy consumers (China, India). At the same time, China and Russia, establishing a long-term partnership, have a preference in the negotiation process for other counterparties, who are given special attention. India, by sources of supply, delimits spheres of influence with China (including in Central and South-East Asia). Brazil has access to Asian markets, producers have sufficient investment. And in aggregate all this forms quite strong positions with regard to the transition to "clean energy" within the entire international community. Given this state of affairs, Russia had every chance to consolidate its position as the largest exporter of energy resources and to take a central position in the BRICS energy markets. In practice, unfortunately, these hopes have not been justified yet, which makes scientists of all branches of science, including legal scholars, think about it. Modern Russian scientists—lawyers are called upon to find adequate legal methods and means for overcoming the current situation. Let's try to analyze it.

In 2009, following the results of the first BRIC summit in Yekaterinburg, it was proposed the introduction of a rather promising formulation on energy, which would include expanding cooperation between producers and consumers, ensuring the security of energy transit, diversifying energy supplies, developing investment in the fuel and energy complex and creating infrastructure. In addition, considerable attention was paid to the prevention of climate change, which prompted calls for the solution of social and economic problems. Subsequently, the emphasis shifted to more traditional for international declarations on the development of renewable energy sources (hereinafter referred to as RES), as well as universal access to energy and improving energy efficiency. In the final declarations of the BRICS summits, the cooperation of the member countries in the field of traditional energy was regulated, but either in the context of preventing excessive volatility of commodity prices, or as a reservation to maintain the dominant role of fossil fuels (Gudkov and Lakhno 2008).

Back in 2010, BRIC countries in the field of energy planned to work together in accordance with national priorities in order to facilitate the use of renewable energy in international cooperation and exchange of experience in the use of renewable energy, including with respect to policies and technologies related to with biofuels (President of Russia 2017). Also, it was announced about cooperation in the field of research and development, training, advisory services and technology transfer in the energy sector (Plakitkin 2013).

Within the framework of promoting sustainable consumption and energy production, which is necessary for the further economic development of the BRICS countries, the following key tasks will contribute to the solution: balance of interests, predictability of supply and demand, transparency. These tasks are in line with the goal of strengthening the energy security of the BRICS countries and are aimed at overcoming the uneven distribution of traditional energy sources and the scarcity of their reserves, and are designed to ensure a significant increase in energy consumption in developing countries and are enshrined in the "BRICS Economic Partnership Strategy", adopted in Ufa on July 9, 2015 (The official website of the Russian Federation presidency in BRICS 2017a).

In this connection, modern projects in the sphere of energy resources supplies are aimed at overcoming territorial and historical disunity and developing foreign trade cooperation. So, in the long term until 2020, the countries plan to start exporting gas to China by a branch from the "Siberia Power" gas pipeline.[2] At the same time, despite the indicated interests of the parties in closer interaction of periodically negotiated negotiations between countries, India still continues to be one of the promising but not yet mastered markets for the sale of energy resources. This state of affairs concerns both the supply of oil and the supply of natural gas.

Energy technologies developed in the BRICS countries can ensure progress in the sustainable development of energy policy, both in the intercountry space in

---

[2]In May 2014, the parties signed a 30-year contract for the supply of gas to China to 38 billion cubic meters of gas per year (*Note by the authors*).

general, and in a single national competence. For example, a long-term partner discussion resulted in the creation of higher added value in the oil and gas chemical industry. In Russia, despite the increase in the production of associated "fatty" gas, it is logical not to burn it, but to process it. For all Russian projects in this area, there is a need for markets and financing. There are also elements of import substitution, but Russian production is too large to focus only on the domestic market.

Prospects for the development of BRICS energy cooperation are real only at low costs and efficient technologies. In the interests of public safety and financial insolvency, many countries had to abandon plans related to the construction of nuclear power plants. The rise in prices in the energy sector once again brought about revival. There was an awareness of the need to establish and legal standardization of more stringent security criteria, as well as unified, more detailed and rigid legal regulation of competition rules in energy markets.

At the present stage, Russian companies are implementing projects in various areas of the fuel and energy complex of the BRICS countries. The largest of them include the construction of nuclear power plants in China and India, in which Rosatom's subsidiary Atomstroyexport participates on the Russian side. In addition, the Russian company Rosneft is involved in the construction of the Oil Refinery and the Petrochemical Company in China. Also, with the participation of Rosneft, preparatory work is underway for the development of oil and gas fields in Brazil (Analytical Center under the Government of the Russian Federation 2017d).

The coordinated activities of the BRICS countries in the global regulation of the fuel and energy sector can have a serious impact on the mechanisms in place in this area. However, the management of the common energy policy by the BRICS countries at the global level is still difficult, due to objective differences in the interests of the countries. The effective development of interstate energy cooperation is directly dependent on investment and technologies introduced into the energy sphere. Therefore, at the present stage of development of international energy cooperation of the BRICS member states, the primary task of lawyers is to create a single legal space. As for the foreign trade turnover of energy resources, priority should first of all be given to the unification of the national civil legislation, as a branch of law regulating private-law trade relations, with a foreign element. The fulfillment of this task will contribute not only to the development of foreign trade turnover, intercountry energy cooperation, but also to strengthening the international legal status of BRICS as a whole, as well as to speeding up the pace of integration processes within the international association.

6   Uniform Legal Support for the Foreign Trade Turnover of the Energy Resources of the BRICS Countries

Let us turn to a more detailed analysis of the unified legal support for the foreign trade turnover of the energy resources of the BRICS countries. Until 2015, within the framework of the BRICS, there were practically no separate agreements on cooperation and trade in the energy sphere. The exception is only the Delhi Declaration, adopted at the Summit on December 28–29, 2012, which consolidated the foundations of energy security in the BRICS countries, as well as in the most general form of the development of energy relations (Ryazanova 2015). The current vectors of economic cooperation in all spheres are defined in the BRICS Economic Partnership Strategy (hereinafter referred to as the Strategy), adopted on the basis of the seventh BRICS Summit in Ufa on July 9, 2015. In this document, the BRICS countries emphasize the importance of promoting energy cooperation, as well as sharing experiences in the areas related to energy production and consumption, energy planning (RF Ministry of Energy 2015). According to the Strategy, the expansion of energy supplies from Russia to the exporting countries is the main direction of development of cooperation in the energy sphere.

Thus, we can state the absence of a unified international legal framework based on the adopted general policy documents. However, in the BRICS summits, the governments of the countries show intentions to improve the systems for regulating foreign economic activities in the energy sphere (The official website of the Russian Federation presidency in BRICS 2017b), representatives of the business community are involved in the negotiations: directors of commercial and investment banks, industrialists, entrepreneurs, which makes it possible to evaluate the proposed legal solutions from the point of view of their practical use for concluding contracts and implementing joint projects.

In addition, the analysis of the legal regulation of energy resources turnover in the BRICS countries showed that the documents elaborated by the international association prefer the diversification of export energy markets based on long-term energy supplies and the development of legal mechanisms for international energy cooperation. They give preference to the forecasting of energy consumption, development of recommendations for the development of energy markets in order to ensure energy security and economic development, exchange of experience and technologies in the field of energy efficiency, renewable energy sources, energy saving and carrying out joint research in the field of energy-saving technologies, energy storage technologies for new and renewable energy sources (Zeng et al. 2017). This is confirmed by the principles, objectives and content of the provisions on the foreign policy activities of the Russian Federation vis-à-vis the BRICS countries, enshrined in the "Concept of the Russian Federation's Participation in the BRICS Association" (hereinafter—the Concept) (The concept of the participation of the Russian Federation in the BRICS union 2013).

At the same time, in the process of forming such legal mechanisms aimed at regulating the foreign trade turnover of the energy resources of the participating countries, they build both a common policy and bilateral relations that do not contradict the national interests of the member countries. Thus, in a joint statement of the Russian Federation and the People's Republic of China "On mutually ben-eficial cooperation and deepening of relations of comprehensive partnership and strategic cooperation" strategic tasks are named. This is the formation of strong relations of Russian-Chinese strategic cooperation in the energy sector; intensifi-cation of cooperation in the field of energy, electricity in the use of alternative energy sources; ensuring through joint efforts the energy security of both countries, integration integration and the world as a whole (Joint Statement of the Russian Federation and the People's Republic of China on Mutually Beneficial Cooperation and Deepening the Relationship of Comprehensive Partnership and Strategic Cooperation 2015; Korshunov 2013).

Thus, most of the unified rules directly regulating the foreign trade relations of the BRICS countries in the field of energy should be concentrated in the national law of the member countries in accordance with the adopted program acts and the technical standards of the BRICS.

For example, according to the Delhi Declaration of the 4th BRICS Summit, energy from fossil fuels still plays a leading role in the energy balance. However, the BRICS countries should not only participate in the expansion of the use of clean and renewable energy sources, as well as energy-saving and alternative technologies to meet the growing needs of economies and peoples, but also in the implementation of this activity in strict compliance with relevant requirements and standards of operational safety (The official website of the Russian Federation presidency in BRICS 2017c).

Following the results of the 2015 summit in Ufa, the declaration significantly expanded the wording with the adoption of the "BRICS Economic Partnership Strategy". Energy is again proclaimed as one of the priorities of interstate integration cooperation, and the main focus is on infrastructure development, expansion of energy supplies and investments in energy projects. At the same time, unfortunately, there is no indication of the need for a coordinated foreign policy in the fuel and energy sphere. However, the basis of the parties' interest is trade and investment activity, as well as the need for technology exchange within the association (The official website of the Russian Federation presidency in the BRICS 2017d).

November 20, 2015, based on the results of the ENES 2015 forum, a Memorandum of Understanding was signed in the field of energy saving and energy efficiency. It is planned to develop a joint list of energy-efficient technologies, enhance cooperation between the public sector, private companies and international development banks (Joint Statement of the Russian Federation 2015). In addition, the subject of discussion was the need to adopt comprehensive cooperation programs in the field of energy conservation and energy efficiency, as well as the creation of a joint energy agency on the basis of the BRICS and the possibility of conducting mutual settlements in national currencies (The official website of the Ministry of Energy of the Russian Federation 2017).

The activities of the BRICS block in the arena of international energy diplomacy are severely limited by the completely different interests of the participants, which affects the speediness of coordination and makes it practically impossible to form a single strategic goal, namely, a new structure for regulating global energy. At the same time, the possibility of uniting does not add much from the point of view of the direct supply of energy resources. After all, the supply and demand could have found each other outside the BRICS format. Nevertheless, we believe that the association can provide new investment opportunities, effective implementation of which will facilitate the unified legal framework within the framework of the interstate association. Thus, the activities of the joint New Development Bank, which in turn mobilizes resources for projects in the field of infrastructure and sustainable development in the BRICS countries, is directed to the investment development of BRICS on a unified regulatory basis (The official Internet portal for legal information 2015). The BRICS format with the participation of the New Development Bank allows the following.

Firstly, due to the possibility of introducing multilateral rules for investment regulation, to introduce an additional degree of investment protection, as well as to minimize threats of politically motivated interference by third countries that previously were intermediaries in the placement of capital. Secondly, it became possible to accumulate funds for projects that are important for several members of the association. At the same time, the possibility of their joint implementation, sharing of results, including new technologies, or obtaining positive external effects is considered.

However, due to the use of new financial organizations and instruments, many management issues arise, the most important of which need to be resolved through the development and harmonization of joint projects, prioritization and distribution of the actual burden on certain members of the BRICS. Therefore, the agenda always raises the question of the fair distribution of investments within the BRICS and the use of a total socio-economic effect for the BRICS countries in general, rather than declarative targets.

Thus, the BRICS summit for its relatively short history has taken a sufficiently weighty position in the world community in the energy, financial and other areas of international cooperation. The main merit of BRICS is the maintenance of a just and democratic multipolar world order, based on the supremacy of international law.

Meanwhile, the difficulties associated with the formation of a unified legal space for the foreign trade turnover of the energy resources of the BRICS countries at the present stage are seen, first of all, in the absence of a legal personality of the new type of integration structure. The formation of the BRICS as a full-fledged international organization is hampered by the lack of formal requirements for this status that distinguish it from other international organizations of a regional type (for example, ASEAN, APEC, SCO, EU, NAFTA) and of an interregional nature, such as the Asia-Europe Meeting, which includes the countries of East Asia and Western Europe (Lakhno 2012).

In addition, there are internal constraints, including several interdependent and interrelated circumstances. First, the natural aspiration of the BRICS participants to preserve their independence as sovereign states, determining independently the priorities of their political, economic, legal and social development. Secondly, the socio-economic state of the BRICS member states, which is difficult to assess under a single standard due to tangible differences in the socio-economic structure, the growth rates of national economies, the liquidity and stability of national currencies, the incompatible level of competitiveness and investment attractiveness of national economies, which causes difficulties in the process of legal standardization. Meanwhile, it is economic relations that are the basis, the first principle and the main prerequisite for the formation of domestic civil-law regulation, which creates a direct legal basis for the foreign trade turnover of the energy resources of the united countries. We believe, that precisely these factors hinder the accelerated development and impede the processes of unification of the legal regulation of foreign trade turnover of energy resources in the civil law regulation of the participating countries.

However, the priority areas of intergovernmental energy cooperation identified within the framework of the BRICS aimed at developing innovation, security, development of the trade system predetermine the need for not only a program and minimum standardization of a general nature, but also the formation of an agreed vector of development of civil, antimonopoly, customs legislation of the member states at the international level (Gudkov and Lakhno 2008).

7 Harmonization of Civil-Law Regulation of Foreign Trade Turnover of BRICS Energy Resources

Analyzing the trends in the progressive development of legal support for modern activities in the energy sector through the creation of the BRICS integration legal field, that the Russian legislator needs to pay attention to the prospects of harmonizing the development of civil, customs, and antimonopoly legislation of the participating countries with a view to strengthening and stabilizing the priority areas of Russia's foreign economic cooperation.

Such harmonization should be focused, first of all, on the relevant rules for the regulation of civil relations within the framework of interstate associations with the participation of the Russian Federation, on the use in the civil legislation of the Russian Federation of the latest positive experience in the modernization of the national civil codes of the BRICS member countries and affect the provisions of the Civil Code of the Russian Federation (Marchukov 2015). The process of formation of a uniform legal regulation of civil law relations complicated by a foreign element in the energy sector in the BRICS member states requires a consistent state stimulation of the creation of unified legal foundations of a common energy policy in framework regulation in the sphere of intercountry turnover of energy resources, which can create the basis for the unification of private-law regulation (Marchukov 2015).

The existing agreements are, in fact, program acts that establish only general directions for the development of cooperation between the participating countries, and do not contain specific provisions on the regulation of the energy market. In this context, it is obvious that it is necessary to modernize the private legal basis for the circulation of energy resources, which includes the unification of national civil and international private law in the sphere of contractual relations, the basis of which is a contract of sale, which will ensure the stability, timeliness and attractiveness of foreign trade partner relations with the Russian Federation for foreign counterparts from the countries participating in the interstate association.

The revision and unification in the aspect of the subject matter under consideration should, first of all, be subject to the provisions of the Civil Code of the Russian Federation regulating foreign trade relations. Since the main legal instrument for the implementation of international trade relations between the commercial structures of the BRICS countries is the foreign economic transaction, primarily the foreign trade agreement.

Foreign trade contracts are the direct basis of foreign economic cooperation. In the framework of the topic under consideration, first of all, contracts on the export or import of energy resources should include the parties to which are entrepreneurs and business structures of different countries (residents and non-residents) participating in the inter-state association BRICS.

It is through the signing and execution of a foreign trade contract for the purchase and sale (supply) of energy resources that most of Russia's foreign trade turnover realized. The countries of the BRICS are not an exception. Therefore, the process of unifying the international legal regulation of foreign trade relations of the BRICS countries should, first of all, touch upon the provisions on foreign trade sale and purchase.

The international standard-legal act designed to unify the regulatory framework for the contract of sale between participants in world trade is the Vienna Convention on Contracts for the International Sale of Goods of 1980, ratified only among Russia, China, Brazil among the BRICS countries. However, these countries have acceded to the convention with disclaimers. The Hague Conventions of 1964: the Convention on Contracts for the International Sale of Goods and the Convention on the Formation of Contracts for the International Sale of Goods have not ratified by the States Parties to the Inter-State Association BRICS. Therefore, for the member states of the BRICS, the provisions of the Vienna Convention of 1980 are of particular importance, which should be taken into consideration by national legislators in the process of convergence of civil regulation.

China acceded to the Vienna Convention in December 1986 with the following reservation: The Vienna Convention on the Contract for the Sale of Goods may be applied only if another State that is the Contracting State of the contractual partner has also ratified this Convention. Thus, at present the application of this multilateral international agreement aimed at unifying the rules of international trade by a party from China to the BRICS countries is possible only with Russia and Brazil (Startsev 2015).

Article 1 of the Vienna Convention provides for the application of the provisions of the Convention to contracts for the sale of goods between parties whose commercial enterprises are located in different States when these States are Contracting States; or when, according to the rules of private international law, the law of a Contracting State is applicable.

China's reservation to Article 1 of the Convention, excluding the application of the Vienna Convention in private law relations with the BRICS countries—India and South Africa and until recently Brazil, which acceded to the Convention on March 4, 2013, complicates the cooperation between these countries.

Therefore, in case of disputes arising out of foreign trade contracts between the parties from China and South Africa or India, the court will have to decide the issue applicable to the contractual obligations of the parties, in the absence of the choice of the right by the parties.

In accordance with civil law, in case one of the parties is a person whose personal law is Russian law, application of the norms of the Vienna Convention is possible. Under India's law, the parties to the treaty are also free to choose the applicable law (unlike Brazil, where the choice falls on the right side of the acceptor) (Agusto 2013). However, if a dispute arising out of such a contract is to be considered in an Indian court, the foreign law will serve only as one of the actual circumstances to be proved, which is due to the fact that for countries of common law (for example, the United States and Great Britain) the qualification of foreign law, as a fact (Timokhov 2004).

Russia, China and Brazil made to the 1980 Convention a reservation to Article 11: "It is not required that the contract of sale be concluded or confirmed in writing or subject to another form requirement. He can be proved by any means, including testimony." Thus, according to the general rule of the Vienna Convention, freedom of form of contract is established. The essence of the reservation is the recognition of an insignificant oral form of the contract of sale. Consequently, the contract of sale concluded by the parties from Russia, China and Brazil can only be written. In India there is a more flexible rule on this issue: the contract of sale can be concluded in any form, similarly in South Africa: the contract of sale can be concluded either verbally or in writing, and also partially orally, in part—in writing (Startsev 2015).

Thus, the legal regulation of purchase and sale in the BRICS is difficult to call unified, but the fact that the majority of the members of the association participate in the Vienna Convention and the possibility of its application in these countries, create the necessary prerequisite necessary for the successful implementation of foreign trade turnover of energy resources of the unified legal space of the integration intercountry union (Rozenberg 2010).

Consider the national legislative framework for foreign trade regulation of the BRICS countries. As for China, liberalization of the foreign trade regime is one of the main directions of transformation of the sphere of legal regulation of foreign trade, after the country's accession to the WTO in 2001. The new version of the 1994 Law of the People's Republic of China "About Foreign Trade" (Laws of the PRC, Legislation and Law of China 2017), which came into force on July 1, 2004, is aimed, inter alia, at enhancing the capabilities of China's national industries to protect interests in foreign trade in goods. The purpose of the Law is to expand "external openness", stimulate the development of foreign trade, maintain the proper order of foreign trade, protect the legitimate rights and interests of entrepreneurs (Article 1) by creating a unified system of foreign trade and encouraging its development (Article 4). The law is consistent with the socialist market economy, stimulates and develops trade relations with other countries and regions on the principles of equality and mutual benefit (art. 5).

When determining the statutory statute of an international commercial transaction, the Chinese conflict of laws law, which arose in the last quarter of the 20th century, proceeds from the "principle of autonomy of the will of the parties". Responding to the requirements of justice and justice, "the principle of closest connection" in Chinese law has a supporting role. All legal systems are equally equal in legal aspect, however in the content aspect they can have serious differences, and the parties have the right to choose a legal system suiting both parties and corresponding to the real need to regulate their contractual relations. We believe, that this inconsistency of approaches is a factor that negatively affects the process of unification (Belikova 2015).

In India, the main law on foreign economic activity is the Act on the Development and Regulation of Foreign Trade, 1992 (Vakilno 2017), which provides for the development and regulation of foreign trade by facilitating the import and increase of exports from India. Proceeding from the provisions of the above-mentioned legal act, the authority of the government (the Ministry of Trade and Industry) includes control and facilitation of foreign trade through the creation of a single system for the promotion and regulation of foreign trade (Costa Lazota 2013).

In South Africa, the main law regulating foreign economic activity is the "Act on Administration of Foreign Economic Activities" of 2002 (WIPO 2017). The law regulates the creation of a single Commission for Foreign Economic Affairs—an independent institution responsible for controlling imports and exports in South Africa. The law provides for the absence of restrictions on private enterprise for foreigners and the same legal conditions for foreign and local investors (Costa Lazota 2013).

In Brazil, in turn, there is still no systematized legislation that determines the basis for state regulation of foreign economic activity. In this sphere, a number of federal laws and by-laws—decrees, resolutions, "provisional measures", circulars, decrees, official communications, normative instructions of various state legislative and executive apparatus (Bezbakh and Puchinsky 2004).

The study found that all law and order provides for the existence of the following three conditions for the existence of a valid contract. First, it is the ability to act (finding common sense, etc.) and the legal capacity of the parties to the contract. Secondly, an indispensable condition is the need for concretization of the subject of the contract, including from the point of view of compliance of the contract itself with the provisions of the current legislation. Thirdly, the presence of free expression of will be expressed in response to an offer in the form of its acceptance. Termination of the treaty in all BRICS countries is also envisaged for the following general reasons: (1) at achievement of the legal purpose on which the agreement of the parties has been directed, that is, at its proper execution; (2) by the will of the parties to the agreement—in the presence of compensation, offset, novation or forgiveness of debt; (3) for objective reasons—with the coincidence of the parties in one person; impossibility of execution; issuance of a state act; death of a citizen or liquidation of a legal entity; (4) when the contracts are not properly executed (for example, if there is a material breach by one of the parties of the terms of the contract). The same grounds for termination of the contract are also applied in contracts with the participation of consumers.

8 Conclusion

Thus, the fact that the majority of the members of the association participate in the Vienna Convention and the possibility of its application in these countries, the uniformity in the national legislative approaches of the BRICS countries regarding the recognition of contracts (deals, agreements) related to the foreign trade turnover of energy resources that define them as prisoners valid), as well as the grounds for their termination create the necessary prerequisite necessary for the successful implementation of foreign trade turnover of energy resources of the unified legal space of the integration intercountry union.

However, it is obvious that the creation of an international unified regulatory legal foreign trade regulation in the energy sector based on adopted general policy documents hinders not only the objectively different interests of the participating countries that have not yet reached a certain degree of harmony in relations, but also the lack of a legal personality of the new type of integration structure, which entails difficulties in observing the guarantees and assumed commitments of the parties to which no counter-measures apply.

In this connection, on the basis of the results of a comparative legal analysis of Russian civil legislation and the legislation of the BRICS countries in the part of unified rules of international trade, first of all, the issues applicable to contractual obligations of the parties, as well as in the absence of the choice of the right by the parties and the establishment of a mandatory form of an international contract of sale and also the difficulties in regulating the obligations of the parties revealed in connection with it we can conclude, that the unification of the legal regulation of foreign trade relations of the BRICS countries in the field of energy at this stage should be concentrated in the national law of the participating countries.

References

Agusto, K. L. (2013). The contract of international sale and purchase: applicable rules on the law of the Russian Federation and Brazil', *Collected reports of graduates of the Master's Degree of the Chair of Civil and Labor Law of the Peoples' Friendship University of Russia*, Publishing house "PFUR", Moscow.

Analytical Center under the Government of the Russian Federation 2017a, *Energy Cooperation Potential of the BRICS Countries*. Retrieved March 23, 2017, from http://ac.gov.ru/files/publication/a/5941.pdf.

Analytical Center under the Government of the Russian Federation 2017b, *Development of oil transportation*. Retrieved January 23, 2017, from http://ac.gov.ru/files/publication/a/9072.pdf.

Analytical Center under the Government of the Russian Federation 2017c, *Development of solar energy*. Retrieved March 23, 2017, from http://ac.gov.ru/files/publication/a/11725.pdf.

Analytical Center under the Government of the Russian Federation 2017d, *Investment in the fuel and energy complex*. Retrieved February 5, 2017, from http://ac.gov.ru/files/publication/a/2992.pdf.

Belikova, K. M. (2015). *National peculiarities and perspectives of unification of private law of the BRICS countries: textbook*, Publishing house "PFUR", Moscow.

Bezbakh, V. V., & Puchinsky, V. K. (2004). *Civil and commercial law of foreign countries: a training manual*. Moscow: ICESM.

Carr, I. (2013). *International trade law* (5th ed.). London: Rutledge.

Costa Lazota, L. A. (2013). Regulation of BRICS foreign trade activities: prospects for convergence of approaches, BRICS countries: trends and development prospects, *Materials of the BRICS youth scientific conference in the modern world: features and perspectives of strategic partnership, Moscow, MGIMO (U), 15 March 2013*, MGIMO (University), Moscow.

Gudkov, I. V., & Lakhno, P. G. (2008). *Actual problems of legal regulation of energy relations*. Moscow: Energy and Law.

Inshakova, A. O., Frolov, D. P., & Ryzhenkov, A Ya. (2015). *Institutional analysis of the nanotechnological "revolution": synthesis of economics and law,*. St. Petersburg: Publishing House "Aleteyya".

Joint Statement of the Russian Federation and the People's Republic of China on Mutually Beneficial Cooperation and Deepening the Relationship of Comprehensive Partnership and Strategic Cooperation [Signed in Moscow on March 22, 2013] 2015, *Problems of the Far East*, no. 3, pp. 4–12.

Korshunov, C.A. (2013). *BRICS Association and Russian-Chinese Relations*, Moscow.

Lahno, P. G. (2012). Energy law in the 21st century: experience of Russia and Foreign countries. *Business Law, 2*, 15–20.

Laws of the PRC, Legislation and Law of China 2017, *Laws of China (PRC) on Foreign Trade*. Retrieved May 19, 2017, from http://law.uglc.ru/trade.htm.

Leal-Arcas, R., Alemany Ríos, J., et al. (2015). The European union and its energy security challenges: Engagement through and with networks. *Contemporary Politics, 21*(3), 273–293.

Marchukov, I. P. (2015) Community of interests of the BRICS countries in the field of energy: The tasks of the common energy policy and energy law, *Bulletin of the VolSU. Series 5. Jurisprudence, 4*(29): 167–174.

News. The economy 2017, *Russia in 2016 increased the export of oil and gas*. Retrieved January 23, 2017, from http://www.vestifinance.ru/articles/81070.

Plakitkin, S. A. (2011). World energy—New frontiers of development. *Effective Anti-Crisis Management, 1*, 44–53.

President of Russia (2017). *Art. 20 Joint statement of the heads of state and government of the countries participating in the II BRIC Summit, 2010*. Retrieved February 5, 2017, from http://www.kremlin.ru/supplement/524.

RF Ministry of Energy (2015). *BRICS Economic Partnership Strategy*. Retrieved November 11, 2015, from http://www.brics.mid.ru/.

RIA official website (2016). *News Putin called for the creation of the BRICS energy agency*. Retrieved December 12, 2016, from https://ria.ru/economy/20161016/1479332011.html.

Rosenberg, M. G. (2010). *International sale of goods: commentary on legal regulation and dispute resolution practice*. Moscow: Statute.

Ryazanova, M. O. (2015). Energy cooperation within the BRICS: Relevance of the issue. *Bulletin of MGIMO, 4*(94), 7–11.

Sevastyanova, T. L. (2013) Efficiency of Russian energy companies. *Law and Management of the 21st Century, 3*, 53–56.

Sharko, S. W. (2011). Geopolitical aspects of BRICS, China in world and regional politics. *History and the Present, 16*, 5–7.

Startsev, D. D. (2015). BRICS countries and the Vienna convention on contracts for the international sale of goods 1980: General application. *Legislation and Economics, 3*, 60–66.

The concept of the participation of the Russian Federation in the BRICS union 2013. *Politics and Economics, 1–2*(51), 39–43.

The official Internet portal for legal information (2015). *The Agreement on the New Development Bank [concluded in Fortaleza on July 15, 2014]*. Retrieved July 07, 2015, from http://www.pravo.gov.ru.

The official website of the Ministry of Energy of the Russian Federation 2017. Retrieved February 04, 2017, from http://minenergo.gov.ru/.

The official website of the Russian Federation presidency in BRICS 2017a, *BRICS Economic Partnership Strategy [Adopted in Ufa on 09/07/2015]*. Retrieved February 05, 2017, from http://brics2015.ru/.

The official website of the Russian Federation presidency in BRICS 2017b, *The Ufa Declaration of the VII BRICS Summit [Adopted in Ufa on 09/07/2015]*. Retrieved February 05, 2017, from http://brics2015.ru/.

The official website of the Russian Federation presidency in BRICS 2017c, *Delhi Declaration of the 4th BRICS Summit [Adopted in New Delhi on 29.03.2012]*. Retrieved February 04, 2017, from http://brics2015.ru/.

The official website of the Russian Federation presidency in the BRICS 2017d, *The BRICS Economic Partnership Strategy [Adopted in Ufa on 09/07/2015]*. Retrieved February 04, 2017, from http://brics2015.ru/.

Timokhov, Y. A. (2004). *Foreign law in judicial practice*. Moscow: Volters Kluver.

Uyanaev, C. B. (2013). Russian-Chinese energy cooperation: Signs of a new "level", China in world and regional politics. *History and the Present, 17*, 277–295.

Vakilno (2017). *The Foreign Trade (Development and Regulation)*. Accessed May 19, 2017, from http://www.vakilno1.com/bareacts/foreigntradeact/foreigntradeact.html.

Vysotsky, V. (2012). The oil potential is very significant. *World Energy, 7*, 74–78.

WIPO (2017). *South Africa International Trade Administration Act 2002*. Retrieved May 19, 2017, from http://www.wipo.int/wipolex/en/details.jsp?id=13345.

Zeng, S., Liu, Y., Liu, C., & Nan, X. (2017). A review of renewable energy investment in the BRICS countries: History, models, problems and solutions. *Renewable and Sustainable Energy Reviews, 74*, 860–872.

Printed by Printforce, the Netherlands